Sven T. Stripp

Molecular Background of Oxygen Sensitivity in [FeFe] hydrogenases

Sven T. Stripp

Molecular Background of Oxygen Sensitivity in [FeFe] hydrogenases

Benefits and Challenges in Enzymatic Hydrogen Production

Südwestdeutscher Verlag für Hochschulschriften

Impressum/Imprint (nur für Deutschland/only for Germany)
Bibliografische Information der Deutschen Nationalbibliothek: Die Deutsche Nationalbibliothek verzeichnet diese Publikation in der Deutschen Nationalbibliografie; detaillierte bibliografische Daten sind im Internet über http://dnb.d-nb.de abrufbar.
Alle in diesem Buch genannten Marken und Produktnamen unterliegen warenzeichen-, marken- oder patentrechtlichem Schutz bzw. sind Warenzeichen oder eingetragene Warenzeichen der jeweiligen Inhaber. Die Wiedergabe von Marken, Produktnamen, Gebrauchsnamen, Handelsnamen, Warenbezeichnungen u.s.w. in diesem Werk berechtigt auch ohne besondere Kennzeichnung nicht zu der Annahme, dass solche Namen im Sinne der Warenzeichen- und Markenschutzgesetzgebung als frei zu betrachten wären und daher von jedermann benutzt werden dürften.

Verlag: Südwestdeutscher Verlag für Hochschulschriften GmbH & Co. KG
Dudweiler Landstr. 99, 66123 Saarbrücken, Deutschland
Telefon +49 681 37 20 271-1, Telefax +49 681 37 20 271-0
Email: info@svh-verlag.de

Approved by: Bochum, Ruhr-Universität, Diss., 2010

Herstellung in Deutschland:
Schaltungsdienst Lange o.H.G., Berlin
Books on Demand GmbH, Norderstedt
Reha GmbH, Saarbrücken
Amazon Distribution GmbH, Leipzig
ISBN: 978-3-8381-2800-9

Imprint (only for USA, GB)
Bibliographic information published by the Deutsche Nationalbibliothek: The Deutsche Nationalbibliothek lists this publication in the Deutsche Nationalbibliografie; detailed bibliographic data are available in the Internet at http://dnb.d-nb.de.
Any brand names and product names mentioned in this book are subject to trademark, brand or patent protection and are trademarks or registered trademarks of their respective holders. The use of brand names, product names, common names, trade names, product descriptions etc. even without a particular marking in this works is in no way to be construed to mean that such names may be regarded as unrestricted in respect of trademark and brand protection legislation and could thus be used by anyone.

Publisher: Südwestdeutscher Verlag für Hochschulschriften GmbH & Co. KG
Dudweiler Landstr. 99, 66123 Saarbrücken, Germany
Phone +49 681 37 20 271-1, Fax +49 681 37 20 271-0
Email: info@svh-verlag.de

Printed in the U.S.A.
Printed in the U.K. by (see last page)
ISBN: 978-3-8381-2800-9

Copyright © 2011 by the author and Südwestdeutscher Verlag für Hochschulschriften GmbH & Co. KG and licensors
All rights reserved. Saarbrücken 2011

TABLE OF CONTENTS

Table of figures

Introduction

1.1	**Hydrogenases catalyse uptake and release of dihydrogen**	**1**
	Hydrogen in enlivened Nature	1
	Convergent evolution gave three classes of hydrogenases	2
	Structural differentiation in [FeFe] hydrogenases	5
	The minimal [FeFe] hydrogenases in green algae	6
	The *C. reinhardtii* hydrogenase serves as a role model in [FeFe] hydrogenase research	7
1.2	**Iron-sulphur proteins**	**8**
	Iron-sulphur clusters mediate single electron transport reactions	8
	Iron-sulphur proteins are found in a variety of specialisations	8
	Iron-sulphur proteins and the problem with oxygen	10

Results

2.1	**Optimized over-expression of [FeFe] hydrogenases with high specific activity in Clostridium acetobutylicum**	**13**
2.2	**Immobilization of the [FeFe] hydrogenase CrHydA1 on a gold electrode: Design of a catalytic surface for the production of molecular hydrogen**	**20**
2.3	**The structure of the active site H-cluster of [FeFe] hydrogenase from green algae Chlamydomonas reinhardtii studied by X-ray absorption spectroscopy**	**28**
2.4	**How oxygen attacks [FeFe] hydrogenases from photosynthetic organisms**	**37**

2.5	Electrochemical kinetic investigations of the reactions of [FeFe] hydrogenase with CO and O_2: Comparing the importance of gas tunnels and active-site electronic/ redox effects	44
2.6	How algae produce hydrogen – News from the photosynthetic hydrogenase	56

Discussion

3.1	Heterologous expression and synthesis of [FeFe] hydrogenases	67
3.2	On the electronic structure of the H-cluster	69
3.3	**Immobilisation of [FeFe] hydrogenases on conductive surfaces**	**70**
	Spectro-electrochemical analysis of hydrogenase films	71
	Direct electrochemistry of hydrogenase films on graphite	71
3.4	**Mechanisms of O_2 inactivation in [FeFe] hydrogenases**	**72**
	[NiFe] and [FeFe] hydrogenases display different levels of O_2 sensitivity	73
	A model for the molecular mechanism of O_2 inactivation	76
	Towards the O_2-tolerant [FeFe] hydrogenase	79

Summary and Acknowledgment

Literature

TABLE OF FIGURES

Figure 1 – The cofactor of [Fe] hydrogenases (Hmd)

Figure 2 – The Ni-Fe cofactor of [NiFe] hydrogenases

Figure 3 – The "H-cluster" of [FeFe] hydrogenases

Figure 4 – Different redox states of the H-cluster

Figure 5 – Iron-sulphur cluster arrangement in CpI

Figure 6 – Crystal structures of and DdH and CpI including [FeS] cluster equipment

Figure 7 – Homology model of the algal [FeFe] hydrogenase CrHydA1

Figure 8 – Iron-sulphur clusters in substrate binding and catalysis

Figure 9 – Position of the cofactors in [Fe] and [FeFe] hydrogenases relative to the protein surface

Figure 10 – Changes upon CO oxidation of the H-cluster as monitored by Fe–Fe distances

Figure 11 – Schematic comparison of the set-up in SEIRAS and graphite based electrochemistry

Figure 12 – Selective access to the active site in [FeFe] hydrogenase CpI

Figure 13 – Comparison of k_{inact} for CO and O_2 and the rate of reactivation after CO inhibition (k_{re-act})

Figure 14 – Stepwise degradation of the H-cluster by superoxide

INTRODUCTION

Hydrogen (H^1) is the most abundant and lightweight element in the universe. The stable form of hydrogen under standard conditions is "dihydrogen" H_2. Dihydrogen is an inert molecule with both nuclei sharing two 1s electrons. The H_2 molecule has a bond enthalpy of 436 kJ mol^{-1} which reflects its chemical robustness. This extraordinary high bonding energy makes H_2 an excellent energy carrier. Its specific enthalpy is four-fold that of coal, and about 2.5-fold that of diesel or natural gas [1]. With just two electrons shared, the intermolecular attraction of nuclei is weak. Thus, H_2 is highly volatile and only trace levels are left in the lower atmosphere [2]. Dihydrogen escaped the stratosphere due to either diffusion [3] or oxidation in the course of the onset of oxygenic photosynthesis [4, 5]. Hydrogen is scarcely found in the H_2 form but as hydride ion (H^-) and proton cation (H^+) in water, Earth's crust, and all kinds of life.

1.1 Hydrogenases catalyse uptake and release of dihydrogen

Hydrogen in enlivened Nature

In anaerobic segments of deep lakes and hot springs the concentration of H_2 is much higher than in the stratosphere. Strict anaerobe bacteria and archaea make use of protons as terminal electron acceptor in anaerobic respiration and fermentation. Oxidation of organic matter and generation of ATP is coupled to reduction of protons that alternatively stand in for O_2. Anaerobic respiration is not the only metabolic pathway that produces H_2. In carboxytrophic bacteria of the *Carboxydothermus* genus [6] oxidation of CO to CO_2 is often coupled to proton reduction [7]. Nitrogen–fixing archaea, proteobacteria (e.g., root nodule bacteria *Rhizobia spec.*), and cyanobacteria (*Nostoc, Anabaena*) release H_2 as a by-product in the reduction of inorganic N_2 to bio-available NH_3 [8-10]. Nitrogen-fixation is catalysed by the nitrogenase complex and demands an additional energy input of 16 equivalents ATP per N_2 and H_2.

Microbial H_2 release is versatile and typically occurs under anaerobic conditions. However, H_2 uptake is found in many micro organisms as well. Knallgas bacteria and affiliated species of the *Desulvovibrio* genus use H_2 as a source of electrons to power their metabolism [11, 12]. Interestingly, the notoriety for anaerobiosis is much less pronounced in organisms relying on H_2 uptake than it is found with H_2 release [13]. The non-standard name "Knallgas bacteria" refers to the fact that bacteria like *Ralstonia eutropha* and *Hydrogenobacter spec.* can live lithoautotrophically on a mixture of H_2 and O_2 [14].

INTRODUCTION

Convergent evolution gave three classes of hydrogenases

Hydrogenases are oxido-reductases (EC 1.12) that catalyse uptake and evolution of H_2 with a variety of redox partners. The reaction hydrogenases perform reads $2\ H^+ + 2\ e^- \leftrightarrow H_2$. While the nitrogenase-based hydrogen metabolism is a physiological specialisation, hydrogenases are ubiquitous in strict and facultative anaerobes, including some unicellular eukaryotes [15]. Three classes evolved independently: [NiFe], [FeFe] and [Fe] hydrogenases. These types of hydrogenase enzymes do not share common ancestors [16]. The mechanistic similarities, however, are striking and although this work will focus on [FeFe] hydrogenases it is worth learning about hydrogen catalysis in [Fe] and [NiFe] hydrogenases as well.

[Fe] hydrogenases (Hmd) (also known as "[FeS] cluster-free hydrogenases") have been described for a number of methanogenic archaea. Originally discovered in *Methanothermobacter marburgensis* [17], the hydrogenase of *Methanocaldococcus jannaschii* has been crystallized just recently [18, 19]. [Fe] hydrogenases catalyse the uptake of H_2 in a binary reaction.

[Fe] hydrogenases incorporate a low-spin iron atom bound to a cysteine residue. Two intrinsic CO ligands coordinate this central metal ion [20]. A 2-pyridinol compound ("FeGP") binds the iron atom at two sites presumably: via pyridinol-N and a formyl carbon side chain [21]. Figure 1 shows a schematic drawing of the [Fe] hydrogenase cofactor arrangement. Due to its octahedral geometry, the central iron atom has a vacant binding site. Dihydrogen is

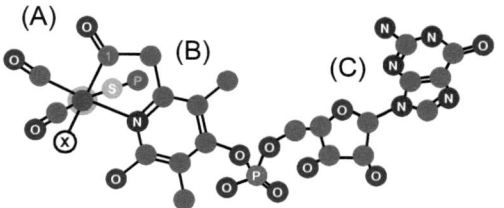

Figure 1 – *The cofactor of [Fe] hydrogenases (Hmd). The central iron atom (A) is coordinated by a single cysteine thiolate and holds two CO groups. The nitrogen atom of the pyridine ring and a formyl carbon atom (1) bind to the iron compound as well. Due to its octahedral geometry, one binding site is vacant (X). The FeGP cofactor is a carboxymethyl-3,5-dimethyl-2-pyridone-4-yl (B) bound to 5'-guanosyl (C) via a phosphate group. The non-iron cofactor methenyl-H_4MPT^+ is not shown (see text).*

dissociated heterolytically: $H_2 \rightarrow H^+ + H^-$. One proton leaves the active site, the hydride is moved to an associated methenyl-tetrahydromethanopterin substrate [22] (methenyl-H_4MPT^+, not shown in Figure 1). The iron atom is *not* catalytically active but serves as a non-redox coordination site, similar to what is found in aconitase (see section **1.2**). [Fe] hydrogenases are sensitive to O_2 damage and efficiently inhibited by extrinsic CO [23].

INTRODUCTION

[NiFe] hydrogenases are the most common hydrogenases in bacteria and archaea [15]. Under physiological conditions, these enzymes are mostly found to catalyse H_2 uptake. Contrary to what has been reported for [Fe] hydrogenases, release of H_2 is possible with [NiFe] hydrogenases *in vitro* as well [24, 25]. A variety of [NiFe] hydrogenase has been crystallized, all of these who stem from sulphate-reducing bacteria [26]. The periplasmatic, membrane-bound hydrogenase of *Desulfovibrio vulgaris* Miyazaki F. is the best-studied member of the "standard" [NiFe] class [27]. No crystal structure was obtained for [NiFe] hydrogenases of the *Ralstonia*-type yet.

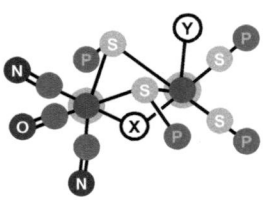

The active site of [NiFe] hydrogenases consists of a nickel atom, sulphur-coordinated by four cysteine residues. Two of these cysteines bridge to another transition metal compound, an iron atom. Nickel exhibits two free coordination sites, one "terminal" and one "bridging". The iron atom has been shown to bind three π–accepting ligands, namely two CN⁻ and one CO group. It shares the bridging coordination site with nickel [26, 28, 29]. Figure 2 displays a schematic drawing of the typical [NiFe] prosthetic group. The complex O_2 and CO sensitivity of this class of hydrogenases is reviewed in the **Discussion**.

Figure 2 – *The Ni-Fe cofactor of [NiFe] hydrogenases. The nickel atom is bound by four cysteine thiolates. Two of these residues bridge to the iron atom. The iron atom carries two CN and as single CO ligand. Furthermore, it shares a free coordination site (X) with nickel. This bridging position is supposed to be the catalytic site. Nickel has another open coordination site (Y) responsible where external CO inhibits the enzyme.*

[FeFe] hydrogenases attracted a lot of attention in recent years due to their ability of rapid H_2 evolution. Matter of fact, [FeFe] hydrogenases are found in H_2 release *in vivo* as well as under laboratory conditions with a turnover number about ten times higher than that of [NiFe] hydrogenases [30]. Two [FeFe] hydrogenases have been characterized by crystallography: "*CpI*" of *Clostridium pasteurianum* and "*DdH*" of *Desulvovibrio desulfuricans* ATCC 5575 [31-33]. Other [FeFe] hydrogenases have been identified, including "*CaHydA*" of *C. acetobutylicum* [34, 35], "*DvH*" of *D. vulgaris [36, 37]*, and "*CrHydA1*", the small algal hydrogenase of *Chlamydomonas reinhardtii* which represents the main subject of this work.

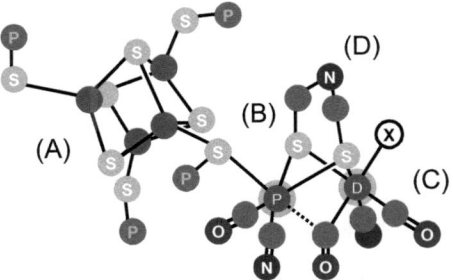

Figure 3 – *The "H-cluster" of [FeFe] hydrogenases. The [4Fe-4S] cluster (A) it anchored to the protein by four cysteine thiolates. One residue (B) brides to the [2Fe-2S] site (C). Relative to this cysteine, a proximal (P) and a distal iron atom (D) are distinguished. Both iron atoms carry at least one CN and CO ligand. Depending on the redox state, another CO can be found bound to the distal iron atom (H_{red}) or in a Fe-Fe bridging position (H_{ox}). Two acid-labile sulphur atoms coordinate the iron atoms. Atop, an azadithiolate ligand has been resolved (D). The distal iron atom exhibits an open coordination site. Hydrogen catalysis, CO inhibition and O_2 inactivation is likely to take place here.*

INTRODUCTION

The prosthetic group of [FeFe] hydrogenases is build from a [4Fe-4S] cluster ("[4Fe]$_H$") and a catalytically active [2Fe-2S] moiety ("[2Fe]$_H$"). This site is commonly referred to as "H-cluster" [31] (Figure 3). The [4Fe]$_H$ subcluster is bound to the protein by four conserved cysteine residues. One of these connects [4Fe]$_H$ and [2Fe]$_H$ – the catalytic part of the H-cluster is covalently attached to the protein by a single thiolate. Inorganic CN$^-$ and CO ligands coordinate the iron atoms of the [2Fe]$_H$ site. Both proximal and distal iron atoms (with respect to the position of the bridging cysteine-S "Fe$_p$" and "Fe$_d$") are substituted with at least one CN$^-$ and CO group [31-33]. Depending on the redox state, another CO is found terminally either at the Fe$_d$ or in a Fe$_p$–Fe$_d$ bridging position [38]. The iron atoms of the [4Fe]$_H$ cluster exhibit a tetrahedral configuration, Fe$_d$ and Fe$_p$ show octahedral geometry. The distal iron atom has a free coordination site that readily attracts hydrogen. A non-protein azadithiolate ligand binds the sulphur atoms of the [2Fe]$_H$ moiety [39].

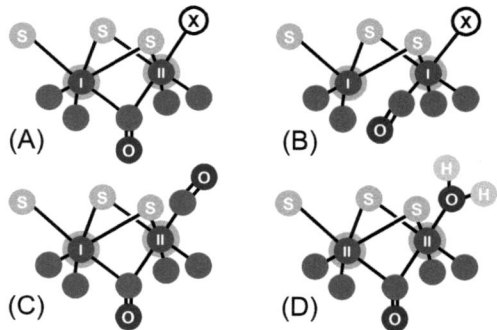

Different redox states have been described for the H-cluster (Figure 4). The oxidised state "**H$_{ox}$**" (A) is paramagnetic and EPR-active. It is commonly denoted as Fe$_p$(I)–Fe$_d$(II) [40]. In the reduced state "**H$_{red}$**" (B), Fe$_d$ is discussed to exhibit a vacant coordination site (Fe$_p$(I)–Fe$_d$(I)–X as in (B)) or bind a hydride anion (Fe$_p$(II)–Fe$_d$(II)–H$^-$, not shown). This state is invisible to EPR. H$_{red}$ has been characterised by infrared spectroscopy due to the missing 1800 cm^{-1} band of the CO bridge which is typically found in H$_{ox}$ [41-43]. External CO binds H$_{ox}$ to give the paramagnetic state "**H$_{ox}$-CO**" (C) [38].

Figure 4 – *Different redox states of the H-cluster. A minimal version of the [2Fe]$_H$ moiety is displayed. The iron atoms are labelled according to their oxidation state. (A) oxidised H$_{ox}$, (B) reduced H$_{red}$ in Fe$_p$(I)–Fe$_d$(I) notation, (C) the CO-inhibited H$_{ox}$-CO, and (D) the inactive form of the H-cluster H$_{inact}$. "X" is a vacant binding site in (A) and (B). Both these states are supposed to be catalytically active. H$_{ox}$-CO is formed from H$_{ox}$ and both states are EPR-active. The overall +2 charge of [4Fe]$_H$ is likely to be unaffected by the redox state. A "super reduced" state has been proposed with [4Fe]$_H^{+1}$ (not shown).*

The O$_2$ sensitivity of [FeFe] hydrogenases is the main subject of this work. Oxygen irreversibly binds Fe$_d$ and leads to inactivation of the [FeFe] hydrogenase. This state "**H$_{ox}$air**" is not to confuse with "**H$_{inact}$**" (D). Some *Desulvovibrio*-type hydrogenases (see next chapter) bind an OH$^-$ or H$_2$O ligand and are stable under air [29]. Catalytic activity is not found in H$_{inact}$ and it takes a reductive treatment to induce hydrogen turnover. In functional analogy to standard-type [NiFe] hydrogenases [44], activated [FeFe] hydrogenases are re-sensitised to O$_2$ damage [45].

INTRODUCTION

The catalytic mechanisms of H_2 turnover remain a matter of debate. [NiFe] hydrogenases were shown to bind a hydride in the Ni–Fe bridging position ("X" in Figure 2) by advanced EPR [46, 47]. From *density function theory* (DFT) calculations, a similar reaction mechanism was proposed for the Fe–Fe binding site in [FeFe] hydrogenases [48, 49]. The H_{red} state of most [FeFe] hydrogenases shows a terminal rather than a bridging CO ligand. Thus, a catalytic Fe–H⁻–Fe coordination might be possible. Another theory involves the azadithiolate nitrogen atom (Figure 4D) in catalysis. From EPR [39] and DFT calculations [50-52] it was suggested that the central nitrogen atom accepts a proton (azadithiolate as a base) while Fe_d binds a hydride ("entatic state" as in Darensbourg et al. [53]). This concerted mechanism is likely to be relevant in H_2 evolution. Protonation of the azadithiolate ligand is further supported by the crystal structures of *Dd*H and *Cp*I: the gas channel directly points to the azadithiolate nitrogen (see Figure 12 in the **Discussion**).

Structural differentiation in [FeFe] hydrogenases

Bacterial [FeFe] hydrogenases are either monomeric (*Clostridia*-type) or dimeric (*Desulvovibrio*-type). With most *Clostrida*-type [FeFe] hydrogenases, a functional bisection is observed. The larger part of the enzyme binds the active site H-cluster and is referred to as "H-domain". The *Clostrida*-type [FeFe] hydrogenases typically holds three [4Fe-4S] and one [2Fe-2S] cluster in the "F-domain". These ferredoxin-type [FeS] clusters are in tunnelling distance to each other and "wire" the active site to the protein surface (Figure 5). The mushroom-shaped structure of *Cp*I (64 kda) clearly illustrates the segmentation of catalytic and accessory domain [31, 33]. In Figure 6B, a cartoon model of the *C. pasteurianum* enzyme is drawn as crystallized. Among the *Clostridia*-type [FeFe] hydrogenases, size and [FeS] cluster composition of the accessory domain varies. While *Cp*I and *Ca*HydA exhibit the maximum number of four clusters, the [FeFe] hydrogenase of *Megasphera elsdenii* holds only two [4Fe-4S] clusters [54].

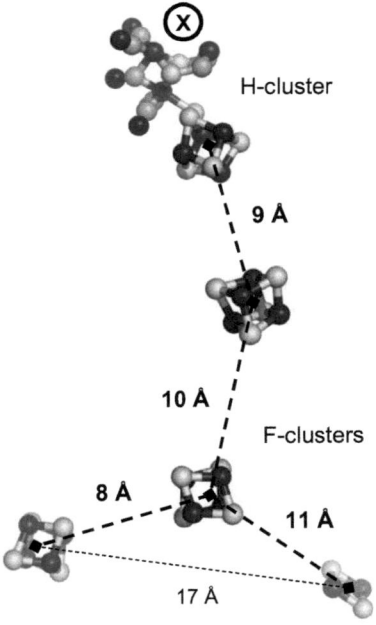

Figure 5 – *Iron-sulphur cluster arrangement in Cp*I. *The F-clusters wire the active site H-cluster to the surface. All clusters are in tunnelling distance (max. ~14 Å). The H-cluster is modelled as* H_{ox} *with a vacant binding site at* Fe_d *(X).*

INTRODUCTION

The dimeric, *Desulvovibrio*-type [FeFe] hydrogenases consist of a large and a small peptide chain. The large subunit of *Dd*H (46 kDa) not only forms the active site but holds two [4Fe-4S] clusters as well. No [FeS] clusters are found with the small subunit (14 kDa). It folds around the large subunit like a belt. Nicolet and co-workers found structural similarities between bacterial ferredoxins and the small subunit of *Dd*H [32]. However, the small subunit is discussed to be relevant in translocation of the protein from the cytoplasm to the periplasmatic space [32, 55]. Figure 6A depicts a cartoon model of the crystal structure of *Dd*H in comparison to the structure of *Cp*I (6B).

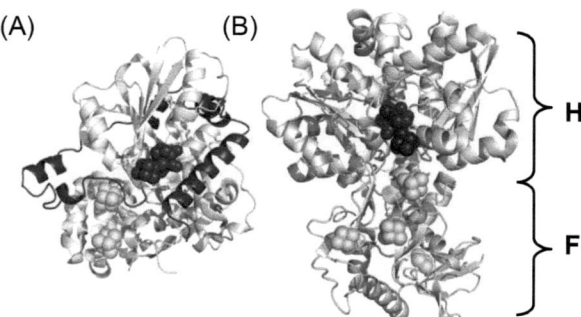

Figure 6 – Crystal structures of DdH and CpI including [FeS] cluster equipment. (A) DdH (1HFE) of Desulvovibrio desulfuricans carries two [4Fe-4S] compounds (yellow spheres) besides the H-cluster (red). These are found with the large subunit (46 kDa). A cluster-free small subunit (14 kDa) folds around the large one like a belt. (B) CpI (3C8Y) of Clostridium pasteurianum consists of a single 64 kDa peptide chain with three [4Fe-4S] and one [2Fe-2S] clusters. This part of the protein is referred to as ferredoxin- or "F-domain" (F). The H-cluster is buried inside the catalytic part of the protein, the "H-domain" (H).

The minimal [FeFe] hydrogenases in green algae

As a representative of Eukaryota, green algae synthesize hydrogenases [56, 57] that have been shown to be of the [FeFe] type and not, as assumed before, [NiFe] hydrogenases [58, 59]. The hydrogenase is usually located in the chloroplast stroma and serves as an "electron valve" under reducing, ATP-deficient conditions – in particular anaerobiosis, which is the key signal for gene expression [59, 60]. Removing electrons from the *photosynthetic electron transport* (PET) chain via proton reduction under conditions where other electron valves like the Calvin Cycle are not available protects the cell from over-reduction. Simultaneously, it allows keeping up the *proton motive force* (PMF) across the thylakoid membrane. Dihydrogen leaves the cell as exhaust. The hydrogenase is coupled to the photosynthetic electron transport chain via photosystem I (PSI) and ferredoxin [60-62]. Numerous authors make use of *C. reinhardtii* in "solar-biological H_2 production" (see Ghiradi et al. [63-66] for review).

The [FeFe] hydrogenases of green algae are known as "*Chlorophyta*-type" hydrogenases [57]. Typically, these enzymes lack the F-domain. No other [FeS] compound than the active site H-cluster is present in *Chlorophyta*-type [FeFe] hydrogenases. Since sequence similarity of the H-domain of *Clostridia*- and *Chlorophyta*-type hydrogenases is fairly high, algal [FeFe] hydrogenase can be viewed as "minimal" version of the typical *Clostridia*-type enzyme [57, 67].

INTRODUCTION

The *C. reinhardtii* hydrogenase serves as a role model in [FeFe] hydrogenase research

Isolated from its native host, the [FeFe] hydrogenase of *C. reinhardtii* is laborious to enrich – one litre cell culture gives only about 60 µg protein [60, 68]. Happe and Naber isolated the enzyme with a H_2 evolution activity of 935 µmol H_2 mg^{-1} min^{-1}. Different systems for the heterologous synthesis emerged since the identity of the "photosynthetic hydrogenase" was resolved [59]. A detailed comparison of the *E. coli*, *C. acetobutylicum*, and *Shewanella oneidensis* systems can be found in the **Discussion** [35, 69, 70].

The biophysical characterisation of *Cr*HydA1 holds important advantages. Due to the absence of accessory [FeS] clusters, iron-specific measurements (*X-ray absorption spectroscopy*, XAS) and redox-sensitive methods (*electron spin resonance*, EPR) are facilitated. In the results section, studies on the *Cr*HydA1 H-cluster by XAS and protein film electrochemistry can be found. Furthermore, O_2 inactivation and maturation of the H-cluster is best studied with *Cr*HydA1 [71-73]. Regarding the first, a tentative interaction of O_2 with F-domain [FeS] clusters does not have to be taken in account. Maturation of *Cr*HydA1 is simplified for the same reason. The catalytic characteristics are solely determined by the H-cluster, accessory clusters do not hamper the analysis.

Although there is no crystal structure of *Cr*HydA1 available yet, theoretical models of the protein structure (Figure 7) help understanding certain aspects of hydrogen turnover. This includes the ongoing issue of gas channels and how these may determine O_2 sensitivity [74-76]. Moreover, the protein–protein interaction of *Cr*HydA1 with ferredoxin PetF was discussed in an elaborate study [67]. Reduction by PetF is crucial because the ferredoxin enzyme is thought to be the central branching point of electrons in the chloroplast stroma [61]. Understanding (and facilitating) the hydrogenase–ferredoxin interaction is another approach to enhance H_2 production with *C. reinhardtii* [63].

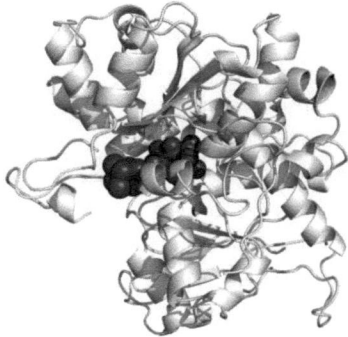

Figure 7 – *Homology model of the algal [FeFe] hydrogenase CrHydA1. The structure was modelled after crystal coordinates of CpI (3C8Y) using the ExPASy Swiss Prot server. CrHydA1 holds no other [FeS] clusters than the H-cluster (spheres).*

INTRODUCTION

1.2 Iron-sulphur proteins

Iron-sulphur clusters mediate single electron transport reactions

Hydrogenases are [FeS] proteins. This vast and heterogeneous protein class characteristically harbours small inorganic [FeS] compounds bound to the protein scaffold by conserved cysteine motifs. Typical [FeS] compounds are [2Fe-2S] and [4Fe-4S] clusters, with two and four equivalents of free sulphur, respectively. Even an iron atom bound by cysteines can be defined as [FeS] site ("Rubredoxin Fe centre" as in [Fe] hydrogenases (Hmd) [20] and as discussed by Edward et al. [77]). These basic modules can be found combined and twisted in a wide range of different [FeS] cofactors [78], including the hydrogen cycling centres of [NiFe] and [FeFe] hydrogenases.

Typically, [FeS] clusters mediate single electron transport reactions. Iron-sulphur clusters accept and donate reducing equivalents with the overall redox state switching between +1 and +2. Depending on the active site protein environment, the range of reducing potentials is widespread. Iron-sulphur clusters have been described from -0.45 to +0.6 V vs. NHE [79, 80]. Common [FeS] sites serve conductive in low-potential reactions (e.g., in hydrogenases) what sets them apart from high-potential cofactors like heme or NAD(P) [81, 82]. Furthermore, it reflects their origin from a reductive, primordial atmosphere [83]. Exception to the rule is the high-potential [4Fe-4S] ferredoxins ("HiPIPs") whose cubane cluster redox state intentionally shuttle between +2 and +3 [84-86].

Iron-sulphur proteins are found in a variety of specialisations

It seems likely that [FeS] proteins were the first catalytic enzymes evolved in the heyday of life [87, 88]. Formation of [FeS] clusters requires ferric iron, thiol and sulphide [89]. Moreover, Venkateswara and co-workers could demonstrate that simple [FeS] compounds form spontaneously and remain stable under reducing and O_2-deficient conditions [90] – typical characteristics of the atmosphere on early Earth. With subsequent oxidation of the atmosphere, the need for protecting protein scaffold of [FeS] clusters grew. Today, [FeS] clusters are "the most abundant and most diversely employed enzymatic cofactors" [83] and a great variety of proteins has been studied.

Ferredoxins are small and soluble redox proteins that are found in bacteria and photosynthetic eukaryotes [91]. The average redox potential is -420 mV vs. NHE. Plant-type ferredoxins harbour a [2Fe-2S] cluster [92] while bacterial-type ferredoxins are equipped with one or two cubane cluster [93, 94]. PetF, a plant-type ferredoxin, receives electrons at the reducing end of the photosynthetic electron transport chain [61]. Due to their trademark role in electron transport, the accessory [FeS] clusters in [NiFe] and [FeFe] hydrogenases have been termed ferredoxin-type clusters.

INTRODUCTION

Rieske proteins are high-potential [FeS] proteins that bind a [2Fe-2S] cluster by two cysteine and two histidine residues [95]. Prominently, Rieske proteins are found as part of the cytochrome b_6/f and cytochrome bc_1 complexes in the chloroplast and mitochondrial membrane, respectively. The redox midpoint potential of the Rieske protein is about 700 mV more positive than that of plant-type ferredoxins [96]. Like all ferredoxins, the Rieske proteins are solely found in electron transport.

The prokaryotic **high-potential [4Fe-4S] ferredoxins** are another ferredoxin-related class of [FeS] proteins. The active site comprises a bacterial-type [4Fe-4S] cluster completely shut-off from the solvent. Therefore, the cofactor in HiPIP can adopt the +3 redox state without oxidative dismissal of an iron atom [85, 86, 97]. Aromatic residues are discussed to determine the extraordinary redox potential [85, 98, 99]. The maximum midpoint potential is +480 mV vs. NHE, even higher than that of the Rieske protein [79].

Aconitase is the best-studied member of the dehydratase family. The enzyme incorporates a [4Fe-4S] cluster that is bound by three cysteine residues (Figure 8A). One iron atom exhibits a free coordination site. Aconitase catalyse the dehydration of citrate to aconitate by binding two hydroxyl groups of (iso-) citrate to this free iron site [100]. Catalysis is mediated by an interplay of certain residues, not electronic participation of the cubane cluster which nevertheless supports substrate binding. Enzymes of the dehydratase-type are widespread [83, 101, 102].

The **radical-SAM enzymes** employ an analogue coordination mechanism to convert aliphatic molecules [103-105]. S-Adenosylmethionine (SAM) binds the free iron site of a catalytic [Fe-4S] cluster via the carboxy and amino groups of the methinone end (Figure 8B). The cubane cluster donates an electron and an adenosyl-radical is formed [106]. Like in dehydratases, the iron coordination sphere changes from tetrahedral to octahedral geometry. However, the [FeS] moiety does not only bind the substrate but displays redox activity as well. The **pyruvate:formate lyase activating enzyme** (PFL-AE) and **biotin synthase** (BioB) are important examples for the radical-SAM superfamily [107, 108]. Maturation of [FeFe] hydrogenases is supposed to be catalysed by radical-SAM proteins [69].

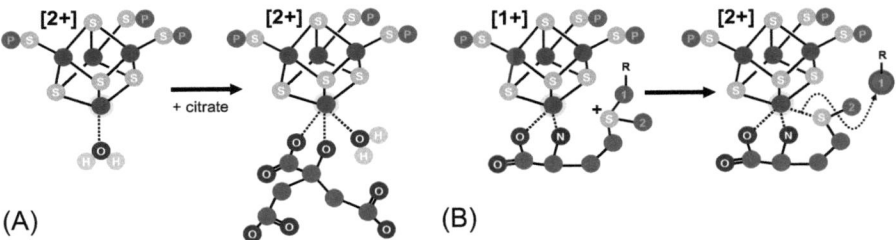

Figure 8 – *Iron-sulphur clusters in substrate binding and catalysis. (A) shows the coordination of citrate to the [4Fe-4S] cluster of a dehydratase In (B) the reaction mechanism of radical-SAM enzymes is displayed. Methionine binds to the free iron site and an electron is passed on to the Adenosyl moiety. The cubane cluster is oxidised 1+ → 2+ in this step. See text for details.*

INTRODUCTION

Next to electron transport and radical synthesis, [FeS] proteins are frequently involved in sensing and regulatory functions. The following chapter will discuss the oxidative disruption of [FeS] clusters. One of the most prominent [FeS] proteins, the fumarate:nitrate reductase, will be introduced subsequently.

Iron-sulphur proteins and the problem with oxygen

Iron-sulphur proteins need to sequester their active site [FeS] clusters from atmospheric exchange if they operate in aerobic environments. Oxygen species like O_2, H_2O_2, and superoxide ($O_2^{\cdot-}$) readily bind transition metal compounds: while iron or nickel are good univalent electron donors, oxygen species accept electrons one-by-one [109]. Oxygen sensitivity is not restricted to *reactive* oxygen species – O_2 is willing redox partner to [FeS] clusters itself [110]. The oxidation state of an oxygen-coordinated [FeS] cluster is +3 ("over-oxidation"). Iron-sulphur clusters as subject to +3 oxidation lose an iron atom due to "solventolysis" as discussed in [97]. The subsequently formed $[3Fe-4S]^{1+}$ compound decomposes if not specifically stabilized by the protein [111-113].

In bacteria and plants, "low-potential" ferredoxins have been reported both sensitive and resistant against oxidative damage [114, 115]. A well-documented example of O_2-stable [FeS] clusters can be found with the chloroplast and mitochondrial inner membranes. The numerous [FeS] proteins of the photosynthetic and mitochondrial electron transport chain (e.g., PSI, NADH dehydrogenase, and the Rieske protein) are insensitive to ambient levels of O_2 [116-118]. The active site in high-potential ferredoxins is lined with aromatic residues. HiPIPs circumvent oxidative damage by excluding any contact of the $[4Fe-4S]^{3+}$ cluster with the solvent. Obviously, O_2 sensitivity of [FeS] proteins is modulated by the peptide scaffold.

Iron-sulphur proteins that mediate electron transport do not necessarily expose their cofactor to the solvent. In [FeS] proteins with substrate binding activity, contact with the solvent is inevitable condition. This selective exposure brings forth the danger of solventolysis [93]. Among the dehydratase and radical-SAM families, different levels of O_2 sensibility have been reported, ranging from insensitive to highly vulnerable [83]. In [NiFe] and [FeFe] hydrogenases, O_2 inactivation is notorious because the enzymes exhibit "tunnels" not only H_2 can travel but O_2 as well [26]. [Fe] hydrogenases display a different strategy. The crystal structure of the O_2-labile [Fe] hydrogenase from *M. jannaschii* shows that the prosthetic group is liberally exposed to the solvent in a cleft formed by the "open" conformation of the protein [19]. Figure 9 shows a structural comparison of the cofactor environment in *Cp*I and in the [Fe] hydrogenase of *M. jannaschii*.

INTRODUCTION

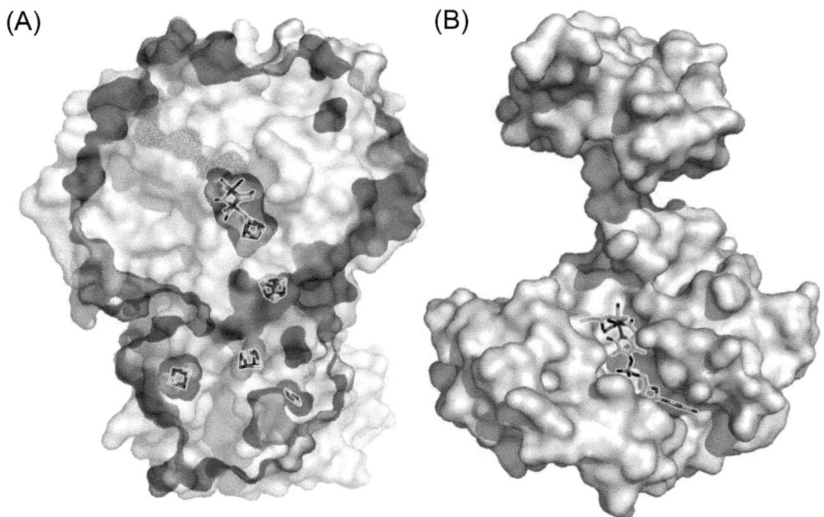

Figure 9 – *Position of the cofactors in [Fe] and [FeFe] hydrogenases relative to the protein surface. (A) View "inside" the [FeFe] hydrogenase CpI (3C8Y). Both H-cluster and ferredoxin-type [FeS] clusters are surround by the protein scaffold. Gas channel "A" is marked as clouds. An intersection is shown. (B) Crystal structure of the [Fe] hydrogenase (Hmd) of M. jannaschii (open conformation)(3F47). The catalytic cofactor Fe-GP is liberally exposed to the solvent.*

There is a wealth of publications reporting the specific reactions of [FeS] proteins with O_2 and π–accepting oxidants like CO, H_2S, and NO [76, 115, 119-121]. Summed up, [FeS] proteins are generally prone to destructive oxidation. Proteins found under aerobic conditions have been evolutionary optimized to function at a specific maximum level of O_2 – typically by shielding the cofactor from solvent. Iron-sulphur proteins that never had to deal with O_2 display pronounced susceptibility. However, with a certain class of oxygen-sensing [FeS] proteins, collapse of a cluster was exploited as regulatory principal [122].

The ubiquitous mammalian [FeS] protein **c-aconitase** holds a cubane cluster with a single uncoordinated iron atom (as discussed above) [123]. This atom is naturally prone to oxidation and the Achilles' heel of the cluster. Upon dismissal, c-aconitase acquires the capability to bind RNA. A multifunctional signal cascade is triggered subsequently [124, 125]. The enzyme without its [FeS] core is referred to as "**iron regulatory protein**" (IRP). The RNA region it binds is specifically termed "**iron responsive elements**" (IREs). The IRE / IRP system is the typical sensing mechanism for the availability of ferric iron (which directly relies on the O_2 content of the cell) [126]. A similar principal is found with the *E. coli* transcriptional regulator **SoxP**. Here, the oxidative breakdown of a $[2Fe-2S]^2$ cluster duplex makes SoxP activate the superoxide response regulon [127, 128]. SoxP senses the concentration of intracellular superoxide and protects the organism by triggering its protection apparatus.

INTRODUCTION

Even more elaborate, the *E. coli* transcription factor **fumarate:nitrate reduction** (FNR) has been discussed to undergo a large structural rearrangement upon O_2 sensing [113, 129, 130]. This iron-sulphur protein is essentially responsible for the switch between respiration and fermentation metabolism [131]. The FNR dimer holds a [4Fe-4S] cluster at each N-terminus and binds DNA at the C-terminal regulatory site. Unlike IRP, FNR loses its DNA-binding capacity once the central $[4Fe-4S]^{2+}$ site has been oxidised to a fairly stable $[2Fe-2S]^{2+}$ cluster. Crack et al. traced cluster breakdown via an $[3Fe-4S]^{1+}$ intermediate and reported *in silico* folding studies of a FNR homology model [113, 132]. Two observations are remarkable: (a) cluster dismissal is accompanied by superoxide release, and (b) the final [2Fe2S] cluster on the FNR monomer is apparently not further affected by O_2 or superoxide.

RESULTS

2.1 Optimized over-expression of [FeFe] hydrogenases with high specific activity in *Clostridium acetobutylicum*

Gregory von Abendroth[1], **Sven T. Stripp**[1], Alexey Silakov[2], Christian Croux[3], Philippe Soucaille[3], Laurence Girbal[3] and Thomas Happe[1,*]

[1] Ruhr-Universität Bochum, Lehrstuhl für Biochemie der Pflanzen, AG Photobiotechnologie, 44780 Bochum, Germany

[2] Max-Planck-Institut für Bioanorganische Chemie, 45470 Mülheim an der Ruhr, Germany

[3] UMR5504, UMR792 Ingénierie des Systèmes Biologiques et des Procédés, CNRS, INRA, INSA, 31400 Toulouse, France

* Corresponding author:
Phone: +49 234 32 27026 ; Fax: +49 234 32 14322
E-mail: thomas.happe@rub.de

http://dx.doi.org/10.1016/j.ijhydene.2008.07.122

RESULTS

Optimized over-expression of [FeFe] hydrogenases with high specific activity in Clostridium acetobutylicum

Gregory von Abendroth[a], Sven Stripp[a], Alexey Silakov[b], Christian Croux[c], Philippe Soucaille[c], Laurence Girbal[c], Thomas Happe[a,*]

[a]Ruhr-Universität Bochum, Lehrstuhl für Biochemie der Pflanzen, AG Photobiotechnologie, 44780 Bochum, Germany
[b]Max-Planck-Institut für Bioanorganische Chemie, 45470 Mülheim an der Ruhr, Germany
[c]UMR5504, UMR792 Ingénierie des Systèmes Biologiques et des Procédés, CNRS, INRA, INSA, 31400 Toulouse, France

ARTICLE INFO

Article history:
Received 27 May 2008
Received in revised form
28 July 2008
Accepted 29 July 2008
Available online 30 September 2008

Keywords:
[FeFe] hydrogenase
Over-expression
Codon usage
Clostridium acetobutylicum
Chlamydomonas reinhardtii

ABSTRACT

It was previously shown that Clostridium acetobutylicum is capable to over-express various [FeFe] hydrogenases although the protein yield was low. In this study we report on doubling the yield of the clostridial hydrogenase by replacing the native gene $hydA1_{Ca}$ with a recombinant one via homologous recombination. The purified protein $HydA1_{Ca}$ shows an unexpected high specific activity (up to 2257 μmol H_2 min^{-1} mg^{-1}) for hydrogen evolution. Furthermore, the highly active green algal hydrogenase $HydA1_{Cr}$ from Chlamydomonas reinhardtii was heterologously expressed in C. acetobutylicum, and purified with increased yield (1 mg protein per liter of cells) and high activity (625 μmol H_2 min^{-1} mg^{-1}). EPR studies demonstrate intact H-clusters for homologously and heterologously expressed [FeFe] hydrogenases in the CO-inhibited oxidized redox state, and prove the high efficiency of the C. acetobutylicum expression system.
© 2008 International Association for Hydrogen Energy. Published by Elsevier Ltd. All rights reserved.

1. Introduction

The production of the energy carrier hydrogen from renewable resources will be a fundamental prerequisite for a sustainable hydrogen-driven economy. A variety of biological systems serve as a likely base for the development of renewable hydrogen-production technologies [1–4]. Photosynthetic systems in particular can directly link the light energy-driven photosynthetic electron chain to proton reduction, as e.g. the green alga Chlamydomonas reinhardtii [5]. The central enzyme involved in proton reduction in C. reinhardtii is the [FeFe] hydrogenase $HydA1_{Cr}$ catalyzing the reversible reduction of protons to molecular hydrogen [6–8].

Hydrogenases are classified according to the metal content of their catalytic center into [NiFe], [FeFe] and [Fe] hydrogenases [9,10]. The first two classes of hydrogenases are involved in the direct evolution and consumption of hydrogen. [FeFe] hydrogenases show higher catalytic activities for hydrogen evolution than [NiFe] hydrogenases [11]. The active site of all known [FeFe] hydrogenases consists of an unique binuclear iron center which is called the H-cluster. It is of major interest to understand the reaction mechanism of this highly efficient iron–sulfur cluster. $HydA1_{Cr}$ from C. reinhardtii is a promising model for better understanding of [FeFe] hydrogenase structure and function as it is one of the smallest known member (48 kDa) of this enzyme family, due to lack of the FeS-cluster containing F-domain [6,12]. With a better

---

* Corresponding author. Present address: Ruhr-Universität Bochum, Lehrstuhl für Biochemie der Pflanzen, AG Photobiotechnologie, ND2/170, 44780 Bochum, Germany. Tel.: +49 234 32 27026; fax: +49 234 32 14322.
E-mail address: thomas.happe@rub.de (T. Happe).
0360-3199/$ – see front matter © 2008 International Association for Hydrogen Energy. Published by Elsevier Ltd. All rights reserved.
doi:10.1016/j.ijhydene.2008.07.122

RESULTS

understanding it could become possible to use [FeFe] hydrogenases in either an artificial hydrogen-production system or in vivo systems including optimized hydrogenases for the renewable production of hydrogen. In this context it will be necessary to over-express highly active [FeFe] hydrogenases in large amounts in order to solve their structure in atomic detail and functionally characterize this important catalyst.

An expression system for [FeFe] hydrogenases had been established in Clostridium acetobutylicum, which achieved high specific activities but rather low protein yield [13]. An alternative system, using Escherichia coli as host, led to high amounts of [FeFe] hydrogenases but comparably low specific activities [14]. In the current report, we present an expression system for C. acetobutylicum that allows for generation of increased amounts of homologously and heterologously synthesized protein with high specific enzyme activities. We demonstrate the high efficiency of this expression system by showing the conservation of the intact H-cluster via EPR-spectroscopy. Based on this work it will now be possible to synthesize large amounts of high quality [FeFe] hydrogenases. This will allow for detailed biophysical analyses of this potent enzyme family leading to an improved description and explanation of their structure and function.

2. Materials and methods

2.1. Growth conditions and maintenance

All C. acetobutylicum strains were grown anaerobically in CGM medium as described previously [15]. For recombinant C. acetobutylicum strains, 40 µg ml^{-1} erythromycin was added to agar plate and liquid media. C. acetobutylicum strains were stored in spore form at $-20\,°C$.

2.2. Fermentation experiments

Batch experiments with C. acetobutylicum ATCC 824 recombinant strains were performed in a 2.5 l Minifors®-bioreactor (Infors, Augsburg, Germany) with a culture volume of 2.0 l on CGM medium and a glucose concentration of 60 g l^{-1} as described earlier [13,15].

2.3. DNA isolation, modification and cell transformation

Total genomic DNA from C. acetobutylicum ATCC 824 was isolated as previously described [16]. Plasmids from recombinant strains of C. acetobutylicum were isolated following the protocol of Girbal et al. [17]. All plasmids were constructed initially in E. coli and then transformed into C. acetobutylicum [16].

DNA restriction and cloning were performed according to standard procedures [18]. Restriction enzymes and T4 DNA ligase were obtained from New England Biolabs (Beverly, MA, USA) and Promega (Madison, WI, USA), respectively, and used according to the manufacturer's instructions. DNA fragments were purified from 2% agarose gels with the GFX® gel band purification kit (GE Healthcare, Buckinghamshire, Great Britain).

PCR amplifications with Pfu polymerase (Promega, Madison, USA), Pwo polymerase (Roche, Meylan, France) and with the expand long template PCR system (Roche, Meylan, France) were performed as described by the manufacturer in a PCR thermocycler mastercycle personal (Eppendorf, Hamburg, Germany). Oligonucleotide synthesis and de novo-gene synthesis were performed by Eurofins MWG (Ebersberg, Germany) and DNA 2.0 (Palo Alto, CA, USA), respectively.

2.4. Plasmids and genetic construction

Construction and cloning of the expression vectors pThydA1$_{Ca}$-C-tag and pThydA1$_{Cr}$-C-tag had been described earlier by Girbal et al. [13]. In order to construct and clone the expression vector pThydA1$_{Cr}$-opt-C-tag, a 5 kbp fragment was excised from the plasmid pThydA1$_{Ca}$-C-tag using the restriction endonucleases BamHI and SmaI. The synthetic gene hydA1$_{Cr}$-opt, containing the green algal hydrogenase gene sequence hydA1$_{Cr}$ with optimized codon usage for expression in C. aceotbutylicum, was excised from the working plasmid pJ10:hydA1$_{Cr}$-opt using BamHI and EcoRV yielding a 1.3 kbp fragment. These two fragments were ligated to construct the expression vector pThydA1$_{Cr}$-opt-C-tag. Within this vector the gene hydA1$_{Cr}$ is under the control of the C. acetobutylicum thiolase promoter and the ribosomal binding site (RBS). The C-terminal Strep tag II sequence was introduced with the 5 kpb fragment from the plasmid pThydA1$_{Ca}$-C-tag. The expression vector pThydA1$_{Cr}$-opt-C-tag^{exp}, containing the codon usage optimized green algal hydrogenase gene together with an exposed C-terminal Strep tag II sequence was constructed in a comparable manner, excising the synthetic gene from the working plasmid pJ10:hydA1$_{Cr}$-opt by using BamHI and ScaI. The linker sequence used to expose the Strep tag II contained 72 bp (5'-GATATCTGGAGTGTAGGAGTTAAACTTTTTGGTGGT GGTAGTGGTGGTGGTAGTGGTGGTGGTAGT-3') leading to a 24 amino acid linker in the synthesized protein.

In order to construct and clone the replicative recombination plasmid pcons(hydA1$_{Ca}$-C-tag) used for homologous over-expression of hydA1$_{Ca}$ with the mutant C. acetobutylicum MGC ΔhydA1$_{Ca}$/hydA1$_{Ca}$-C-tag, two homologous regions (HR) had to be amplified. The homologous region upstream of the native gene hydA1$_{Ca}$, was amplified using the primers HydA1 (5'-AAAGGATCCCTTATTAGTATTGATTTAATTAGCTTGCCAT TGTGC-3') and HydA2 (5'-GGGGAGGCCTAAAAAGGGGGATC GATAAAAATTAACGTTTAATCAACGTAAATATTACGTAC-3') yielding a 912 bp fragment, HR1, and introducing a BamHI-restriction site at the 5'-end and a ClaI-StuI-restriction site at the 3'-end. The downstream homologous region was amplified using primers HydA3 (5'-AAAATCGATCCCCCTTTTT AGGCCTCCCCTCTAAGTTGAGGCACATTTATTTTACTATTTTA CTCC-3') and HydA4 (5'-AAAGGATCCGTCTTCTAAATTAAAT ATAGATAATATAAAGTTTCTTTACTTAACACC-3') yielding a 1010 bp fragment, HR2, and introducing a ClaI-StuI-restriction site at the 5'-end and a BamHI-restriction site at the 3'-end. Fusion-PCR using both fragments as a template together with the primers HydA1 and HydA4 resulted into a combined fragment HR1:HR2 through the homologous ClaI-StuI-region with a size of 1922 bp. Digestion with ClaI and StuI allowed the directed insertion of a ClaI–hydA1$_{Ca}$-C-tag-StuI-fragment, which was previously PCR-amplified using the primers HydA5 (5'-AAAATCGATTCCTACATTTTGGGAGGA-TAAACATGG-3') and HydA6 (5'-AAAAGGCCTTGACCATGAT-TACGAATTCTATGAGTC-3'). Subsequent digestion with StuI allowed the insertion of the StuI-MLSr-cassette, which was obtained by StuI-digestion of the plasmid puc18-FRT-MLSR-2 [19] and contained the erythromycin resistance gene mls'. The final construct BamHI-HR1-hydA1$_{Ca}$-C-tag-MLSr-HR2-BamHI was introduced into the BamHI-digested plasmid pcons2-1 [19] yielding the final recombination vector pcons(hydA1$_{Ca}$-C-tag).

2.5. Homologous recombination and screening

The replicative recombination plasmid pcons(hydA1$_{Ca}$-C-tag) was transformed into C. acetobutylicum MGC as described previously [16] and screened for positive recombinants as described within the patent of Soucaille et al. [19] for the positive mutant C. acetobutylicum MGC ΔhydA1$_{Ca}$/hydA1$_{Ca}$-C-tag. Analysis of the positive recombinant was done by Southern blot analysis [20] using 5 μg genomic DNA, digested with HindIII and separated on a 1% agarose gel. The 517 bp probe against hydA1$_{Ca}$ was isolated from the plasmid pcons(hydA1$_{Ca}$-C-tag) by FokI-digestion. The AlkPhos® direct kit (GE Healthcare, Buckinghamshire, Great Britain) was used for labeling the probe with alkaline phosphatase. Detection occurred by chemiluminescence with the CDP-Star-reagent (Roche, Grenzach-Wyhlen, Germany) and visualization using a Luminometer model FlurChem 8800 (Alpha Innotech, San Leandro, CA, USA).

2.6. Purification of Strep tag II – tagged [FeFe] hydrogenases

Protein purification occurred under strictly anoxic conditions as described by Girbal et al. [13]. An optimized one-step purification protocol was established for the heterologously expressed [FeFe] hydrogenase HydA1$_{Cr}$-C-tag^{exp}. In this case, affinity chromatography on a 1-ml Strep-Tactin Superflow® (IBA, Göttingen, Germany) column was carried out using 100 mM Tris-HCl, 0.1 M NaCl, pH 8.0 as buffer, 2 mM dithionite and 2 mM DTT as reducing agents and 2.5 mM desthiobiotin for elution of the protein. The presence of Strep tag II-tagged hydrogenases in the purification fractions was analyzed by hydrogenase activity assays (see below) and immunoblotting after 12% SDS/polyacrylamide gel electrophoresis using Strep-Tactin HRP conjugate (IBA, Göttingen, Germany) at 1:1000 dilution. Low-range standard proteins (GE Healthcare, Buckinghamshire, Great Britain) and pre-stained low-range SDS-PAGE standards (Biorad, Munich, Germany) were used.

2.7. Hydrogenase activity assays

Hydrogenase activity was analyzed by measuring hydrogen evolution as previously described by Winkler et al. [21], except for using 100 mM dithionite and 1-50 μl of protein solutions (max. 50 nM) instead of algal cultures. For determination of Michaelis Menten kinetics, the concentration of the artificial electron donor methyl viologen was varied from 0.1 mM to 40 mM. Enzyme kinetics using the natural electron donor-type protein ferredoxin were performed using 1 mM dithionite and ferredoxin concentrations ranging from 0.05 μM to 40 μM for green algal PetF and 10 μM to 100 μM for bacterial ferredoxin from Clostridium pasteurianum. The values of kinetic constants were determined from double reciprocal plots.

2.8. EPR spectroscopy

In order to analyze the functionality of the H-cluster of the purified [FeFe] hydrogenases, pulse EPR-spectroscopy was applied essentially as described previously [22,23]. The sample preparations were performed under the previously described anoxic conditions. Protein samples were concentrated in Millipore Centricon® Centrifugal Filter Units (molecular weight cut-off 10 kDa) by centrifugation to a final concentration of 10–100 μM. Treatment of samples with CO gas was performed outside the glove box, using gas tight SubaSeal tubes. After the treatment samples were transferred into EPR tubes and frozen in liquid nitrogen. EPR spectra were obtained at Q-band frequencies, using the 2 pulse electron spin echo detected EPR technique [24,25]. Electron spin echo (ESE) after two microwave pulses ($\pi/2$ and π) was detected as a function of the external magnetic field. The delay between the MW pulses was fixed to $\tau = 360$ ns. The length of the $\pi/2$ MW pulse was set to 36 ns and the π pulse to 68 ns. All pulse Q-band EPR measurements were performed on a Bruker ELEXSYS E580 Q-band spectrometer with a SuperQ-FT microwave bridge, working at 33.88 GHz and a temperature of 20 K.

3. Results

3.1. Homologous over-expression of hydA1$_{Ca}$ in C. acetobutylicum

Previously, homologous expression of the clostridial [FeFe] hydrogenase HydA1$_{Ca}$ was possible with high specific activities but protein yield (0.4 mg l^{-1}) was too low for extended biophysical examinations [13]. The newly established optimized method for homologous recombination in C. acetobutylicum [19] gave rise to the approach of engineering an optimal strain for the homologous over-expression of hydA1$_{Ca}$. This method of homologous recombination was applied within the current study for the first time in the purpose of gene replacement in C. acetobutylicum. This way it was possible to design the over-expression mutant C. acetobutylicum MGC ΔhydA1$_{Ca}$/hydA1$_{Ca}$-C-tag. In this mutant the native gene hydA1$_{Ca}$ was replaced via a double cross-over event with the recombinant gene hydA1$_{Ca}$-C-tag, which is equivalent to the native [FeFe] hydrogenase gene except for a C-terminal Strep tag II sequence. The gene replacement was proven by Southern blot analysis (Fig. 1). With this mutant it was possible to obtain a protein yield for HydA1$_{Ca}$-C-tag of 0.8 mg l^{-1}. The gene replacement mutant solely expressed the recombinant hydrogenase hydA1$_{Ca}$-C-tag and no native hydA1$_{Ca}$ (data not shown).

Analyzing enzyme kinetics for HydA1$_{Ca}$ proton reduction using the same in vitro-test conditions as for HydA1$_{Cr}$ revealed a high specific enzyme activity for HydA1$_{Ca}$ of 1750 μmol H$_2$ min^{-1} mg^{-1} (using methyl viologen as electron donor) and 2257 μmol H$_2$ min^{-1} mg^{-1} (using [4Fe4S] ferredoxin from Clostridium pasteurianum as electron donor) (Table 1). EPR-spectroscopic analysis of the CO-inhibited redox state of HydA1$_{Ca}$

RESULTS

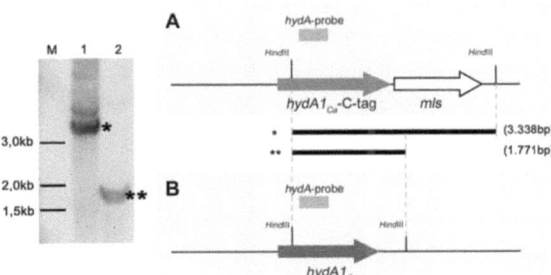

Fig. 1 – Southern blot of HindIII-cut genomic DNA of C. acetobutylicum ΔhydA1$_{Ca}$/hydA1$_{Ca}$-C-tag and C. acetobutylicum ATCC 824 labeled with a hydA-probe and descriptive scheme. HindIII-digested genomic wild type DNA shows the restriction sites as marked in (B) and reveals a band at 1.8 kb within the Southern blot (lane 2, **). The genomic DNA of the mutant C. acetobutylicum ΔhydA1$_{Ca}$/hydA1$_{Ca}$-C-tag shows a band at 3.3 kb within the Southern blot (lane 1, *) due to the C-terminal Strep tag II sequence and the integrated mls-cassette of 1.5 kb in between the two HindIII restriction sites (A). Abbreviations: M, marker; mls, resistance cassette against erythromycin.

revealed the characteristic axial EPR-spectrum and g-values ($g = 2.075, 2.009, 2.009$) (Fig. 2B).

3.2. Heterologous over-expression of the green algal [FeFe] hydrogenase hydA1$_{Cr}$ in C. acetobutylicum

Similar to the homologous expression system mentioned above the heterologous expression of green algal [FeFe] hydrogenases in C. acetobutylicum had been shown with high specific activities but protein yield (0.1 mg l^{-1}) too low for biophysical examinations [13]. Depending on the age of the mutant spores the protein yield was even further diminished to 0.03 mg l^{-1} of cell culture using this expression system (data not shown). Furthermore, there was significant contamination of purified HydA1$_{Cr}$-C-tag with biotinylated pyruvate carboxylase (PYCA, 127 kD) from C. acetobutylicum. The codon usage of hydA1$_{Cr}$ revealed fundamental differences in comparison to C. acetobutylicum resulting in an increased requirement of rare tRNAs and therefore very likely physiological stress for the expression mutant [26]. Adaptation of the codon usage by changing 83% of the codons within the synthetic gene hydA1$_{Cr}$-opt, which was cloned into pThydA1$_{Cr}$-opt-C-tag as previously described for pThydA1$_{Cr}$-C-tag [13], allowed for the reliable expression of significantly increased amounts of algal [FeFe] hydrogenase HydA1$_{Cr}$-C-tag (0.3 mg l^{-1}). PYCA contamination was not detected anymore after expression of the optimized gene hydA1$_{Cr}$ (Fig. 3A).

Using the strain C. acetobutylicum pThydA1opt-C-tag^{exp}, which synthesizes HydA1$_{Cr}$-C-tag^{exp} with an exposed C-terminal Strep tag II sequence resulted in further increase of protein yield up to 1 mg l^{-1} (Fig. 3B). Enzyme kinetics for heterologously expressed HydA1$_{Cr}$-C-tag^{exp} revealed a specific proton reduction activity of 625 μmol H$_2$ min^{-1} mg^{-1} (Table 1).

In order to prove the functionality of the bimetallic active site of heterologously expressed [FeFe] hydrogenases EPR-spectroscopic analysis of the CO-inhibited enzymes was conducted. The CO-inhibited redox state reveals typical axial spectra for H-cluster containing [FeFe] hydrogenases in general [23,27]. It was possible to assign the axial EPR-spectrum with principal g-values ($g = 2.050, 2.008, 2.008$) to heterologously synthesized HydA1$_{Cr}$-C-tag^{exp} (Fig. 2A).

4. Discussion

4.1. Homologously over-expressed [FeFe] hydrogenase from C. acetobutylicum

Gene replacement as a method for creating efficient expression mutants in C. acetobutylicum allowed for deletion of the native hydrogenase gene and incorporation of the recombinant hydrogenase gene within the chromosomic region of hydA1$_{Ca}$. These two events may have caused the twofold increase of purified recombinant hydrogenase in comparison

Table 1 – Kinetic parameters of HydA1$_{Cr}$ and HydA1$_{Ca}$ for methyl viologen and ferredoxin as electron donors						
Enzyme	Electron donor	K_m (M)	V_{max} (μmol min^{-1} mg^{-1})	k_{cat} (s^{-1})	k_{cat}/K_m (M^{-1} s^{-1})
HydA1$_{Cr}$	Methyl viologen	938×10^{-6}	625	535.4	5.7×10^5
	[2Fe2S] ferredoxin	3.4×10^{-6}	526	450.6	1.3×10^8
HydA1$_{Ca}$	Methyl viologen	603×10^{-6}	1750	1951	3.2×10^6
	[4Fe4S] ferredoxin	3.3×10^{-6}	2257	2516	7.6×10^8

RESULTS

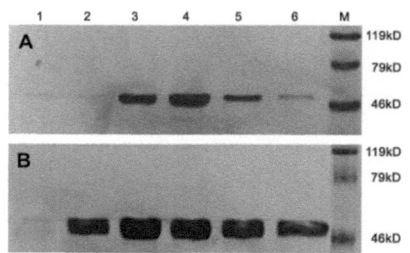

Fig. 3 – SDS-PAGE of purified HydA1$_{Cr}$, coomassie stained, using strain C. acetobutylicum pThydA1$_{Cr}$-opt-C-tag (A) and C. acetobutylicum pThydA1$_{Cr}$-opt-C-tag^{exp} (B). Abbreviations: E1-E6, eluate fractions; M, marker.

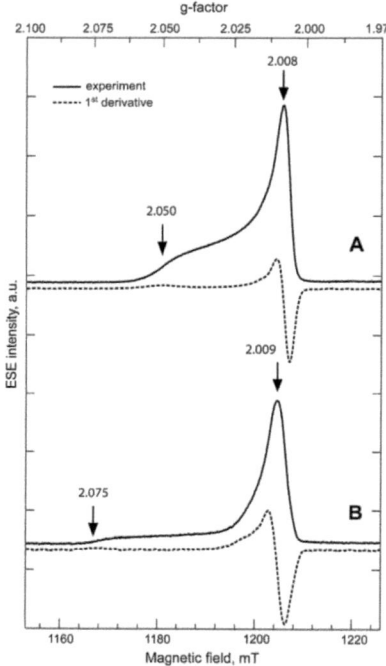

Fig. 2 – Q-band 2 pulse ESE detected EPR spectra of the CO-inhibited states of HydA1$_{Cr}$-C-tag^{exp} (A) and HydA1$_{Ca}$-C-tag (B) measured at 20 K and 33.88 GHz. Abbreviations: ESE, electron spin echo; EPR, electron para resonance.

to the plasmid based expression system described by Girbal et al. [13]. Furthermore, the purified enzyme had a specific activity for hydrogen evolution 10- to 175-fold higher than reported earlier for the clostridial hydrogenase HydA1$_{Ca}$ [13,28]. Compared to the E. coli-based heterologous expression of hydA1$_{Ca}$ the activity described in this paper was more than twenty times higher [14]. The specific activity of HydA1$_{Ca}$ reported in this publication was in the same range as shown for the rather homologous enzyme CpI from C. pasteurianum or the [FeFe] hydrogenases from Megasphaera elsdenii and Desulfovibrio vulgaris [29].

The EPR-spectroscopic examinations of HydA1$_{Ca}$ revealed a spectrum and g-values ($g = 2.075$, 2.009, 2.009) different from those of HydA1$_{Cr}$, but rather similar to the ones found for CO-inhibited CpI ($g = 2.072$, 2.006, 2.006) [30]. Compared to HydA1$_{Cr}$ from C. reinhardtii, the left shoulder of the EPR-spectrum for the clostridial hydrogenases was shifted towards lower values, which may indicate differences in structure and functionality between the observed enzymes.

The method of homologous recombination in the context of [FeFe] hydrogenase gene expression opens new perspectives for the use of C. acetobutylicum as a valuable expression host. Subsequent gene deletions, which are possible with the recombination system applied in this work [19], and multiple gene exchanges could allow for the design of further optimized expression mutants. Furthermore, [FeFe] hydrogenases from a wide range of physiological and evolutionary distinct organisms might be synthesized in a stable and highly efficient manner.

4.2. Heterologously over-expressed [FeFe] hydrogenases in C. acetobutylicum

Heterologously expressed [FeFe] hydrogenases from different green algae revealed similar high specific hydrogen evolution activities compared to the native systems [13]. In contrast, heterologous expression of hydA1$_{Cr}$ in E. coli did not lead to any active enzyme, unless the maturation factors hydEF and hydG were co-expressed. Nevertheless, the enzyme synthesized via co-expression of the maturation factors showed a specific hydrogen evolution activity of only 16% and 21% compared to the native [FeFe] hydrogenase HydA1$_{Cr}$ and the heterologously expressed HydA1$_{Cr}$-C-tag in C. acetobutylicum, respectively [13,14]. For the functional expression of fully active [FeFe] hydrogenases in E. coli additional maturation factors might be necessary, which are presumably naturally available in the expression host C. acetobutylicum. Consequently, the expression of hydA1$_{Cr}$-C-tag^{exp} in C. acetobutylicum revealed a similarly high specific proton reduction activity of 625 μmol H$_2$ min^{-1} mg^{-1} (Table 1), which is about 67% of the native enzyme [6] and 86% of the heterologously expressed enzyme without exposed Strep tag II sequence [13]. The mere insertion of a Strep tag II and a linker sequence had a slightly negative effect on the specific activity of heterologously expressed [FeFe] hydrogenases from C. reinhardtii. Nevertheless, it has to be noted that the specific activity of HydA1$_{Cr}$-C-tag^{exp} reported in this publication was in the same high range as reported earlier for heterologously expressed HydA1$_{Cr}$ [13] and that to our knowledge in this regard C. acetobutylicum is the most suitable expression host for [FeFe] hydrogenases.

The EPR-spectrum of heterologously synthesized HydA1$_{Cr}$-C-tag^{exp} was very similar to natively synthesized and isolated HydA1$_{Cr}$ ($g =$ 2.052, 2.007, 2.007) [23]. This result confirmed that the maturation of the characteristic H-cluster must occur similarly within the expression systems C. acetobutylicum and C. reinhardtii, allowing the heterologous expression of algal [FeFe] hydrogenase with a fully intact H-cluster.

In conclusion, it is now possible to significantly increase the protein yield and maintain a high degree of enzyme functionality within the optimized heterologous expression system for [FeFe] hydrogenases in C. acetobutylicum. This was achieved mainly by adjusting the codon usage and exposing the Strep tag II sequence in the case of green algal hydA1$_{Cr}$. This work has established a reliable method for providing sufficient amounts of enzyme for the biophysical characterization of this smallest known member of the [FeFe] hydrogenase family. This will soon allow a profound insight into the structure and function of this unique class of enzymes.

Acknowledgments

This research work was supported by Süd-Chemie AG, Stiftung der Deutschen Wirtschaft and the Marie Curie fellowship. This work was further supported by the European Commission (7th FP, NEST STRP SOLAR-H2 contract 212508) and the Deutsche Forschungsgemeinschaft (SFB 480).

REFERENCES

[1] Melis A, Melnicki MR. Integrated biological hydrogen production. Int J Hydrogen Energy 2006;31:1563–73.
[2] Ust'ak S, Havrland B, Munoz JOJ, Fernandez EC, Lachman J. Experimental verification of various methods for biological hydrogen production. Int J Hydrogen Energy 2007;32(12):1736–41.
[3] Lin P, Whang L, Wu Y, Ren W, Hsiao C, Li S, et al. Biological hydrogen production of the genus Clostridium: metabolic study and mathematical model simulation. Int J Hydrogen Energy 2007;32(12):1728–35.
[4] Manish S, Banerjee R. Comparison of biohydrogen production processes. Int J Hydrogen Energy 2008;33(1):279–86.
[5] Melis A. Green alga hydrogen production: progress, challenges and prospects. Int J Hydrogen Energy 2002;27:1217–28.
[6] Happe T, Naber JD. Isolation, characterization and N-terminal amino acid sequence of hydrogenase from the green alga Chlamydomonas reinhardtii. Eur J Biochem 1993;214(2):475–81.
[7] Happe T, Kaminski A. Differential regulation of the Fe-hydrogenase during anaerobic adaptation in the green alga Chlamydomonas reinhardtii. Eur J Biochem 2002;269(3):1022–32.
[8] Forestier M, King P, Zhang L, Posewitz M, Schwarzer S, Happe T, et al. Expression of two [Fe]-hydrogenases in Chlamydomonas reinhardtii under anaerobic conditions. Eur J Biochem 2003;270(13):2750–8.
[9] Shima S, Thauer RK. A third type of hydrogenase catalyzing H$_2$ activation. Chem Rec 2007;7(1):37–46.
[10] Vignais PM, Billoud B. Occurrence, classification, and biological function of hydrogenases: an overview. Chem Rev 2007;107(10):4206–72.
[11] Frey M. Hydrogenases: hydrogen-activating enzymes. Chembiochem 2002;3(2–3):153–60.
[12] Winkler M, Hemschemeier A, Gotor C, Melis A, Happe T. [Fe]-hydrogenases in green algae: photo-fermentation and hydrogen evolution under sulfur deprivation. Int J Hydrogen Energy 2002;27:1431–9.
[13] Girbal L, von Abendroth G, Winkler M, Benton PM, Meynial-Salles I, Croux C, et al. Homologous and heterologous overexpression in Clostridium acetobutylicum and characterization of purified clostridial and algal Fe-only hydrogenases with high specific activities. Appl Environ Microbiol 2005;71(5):2777–81.
[14] King PW, Posewitz MC, Ghirardi ML, Seibert M. Functional studies of [FeFe] hydrogenase maturation in an Escherichia coli biosynthetic system. J Bacteriol 2006;188(6):2163–72.
[15] Wiesenborn DP, Rudolph FB, Papoutsakis ET. Thiolase from Clostridium acetobutylicum ATCC 824 and its role in the synthesis of acids and solvents. Appl Environ Microbiol 1988;54(11):2717–22.
[16] Mermelstein LD, Welker NE, Bennett GN, Papoutsakis ET. Expression of cloned homologous fermentative genes in Clostridium acetobutylicum ATCC 824. Biotechnology 1992;10(2):190–5.
[17] Girbal L, Mortier-Barriere I, Raynaud F, Rouanet C, Croux C, Soucaille P. Development of a sensitive gene expression reporter system and an inducible promoter-repressor system for Clostridium acetobutylicum. Appl Environ Microbiol 2003;69(8):4985–8.
[18] Sambrook J, Fritsch EF, Maniatis T. Molecular cloning: a laboratory manual. Cold Spring Harbour Laboratory Press; 1989.
[19] Soucaille P, Figge R, Croux C. Process of chromosomal integration and DNA sequence replacement in Clostridia. Dépôt PCT n° PCT/EP2006/066997: France; 2006.
[20] Southern EM. Detection of specific sequences among DNA fragments separated by gel electrophoresis. J Mol Biol 1975;98(3):503–17.
[21] Winkler M, Maeurer C, Hemschemeier A, Happe T. The isolation of green algal strains with outstanding H2-productivity. Elsevier; 2004.
[22] Silakov A, Reijerse EJ, Albracht SP, Hatchikian EC, Lubitz W. The electronic structure of the H-cluster of the [FeFe]-hydrogenase from Desulfovibrio desulfuricans: a Q-band 57Fe-ENDOR and HYSCORE study. J Am Chem Soc 2007;129(37):11447–58.
[23] Kamp C, Silakov A, Lubitz W, Happe T. Isolation and first EPR characterization of the [FeFe]-hydrogenases from green algae. Biochim Biophys Acta 2008;1777(5):410–6.
[24] Hahn EL. Spin echoes. Phys Rev 1950;80(4):580–94.
[25] Schweiger A, Jeschke G. Principles of pulse electron paramagnetic resonance. Oxford University Press; 2001.
[26] von Abendroth G. Überexpression und Aufreinigung von [FeFe]-Hydrogenasen im Expressionssystem Clostridium acetobutylicum und funktionelle Untersuchung. Bochum: Department for Plantbiochemistry, AG Photobiotechnology; 2007.
[27] Albracht SP, Roseboom W, Hatchikian EC. The active site of the [FeFe]-hydrogenase from Desulfovibrio desulfuricans. I. Light sensitivity and magnetic hyperfine interactions as observed by electron paramagnetic resonance. J Biol Inorg Chem 2006;11(1):88–101.
[28] Demuez M, Cournac L, Guerrini O, Soucaille P, Girbal L. Complete activity profile of Clostridium acetobutylicum [FeFe]-hydrogenase and kinetic parameters for endogenous redox partners. FEMS Microbiol Lett 2007;275(1):113–21.
[29] Adams MWW. The structure and mechanism of iron-hydrogenases. Biochimica et Biophysica Acta 1990;1020:115–45.
[30] Bennett B, Lemon BJ, Peters JW. Reversible carbon monoxide binding and inhibition at the active site of the Fe-only hydrogenase. Biochemistry 2000;39(25):7455–60.

RESULTS

2.2 Immobilization of the [FeFe] hydrogenase *Cr*HydA1 on a gold electrode: Design of a catalytic surface for the production of molecular hydrogen

Henning Krassen[a,§] & **Sven T. Stripp**[b,§], Gregory von Abendroth[b], Kenichi Ataka[a], Thomas Happe[b], Joachim Heberle[a,*]

[a] Bielefeld University, Department of Chemistry, D-33615 Bielefeld, Germany
[b] Ruhr University Bochum, Department of Biochemistry of Plants, SG Photobiotechnology, D-44801 Bochum, Germany

[§] These authors contributed equally to this work.

*Corresponding author:
Phone: +49-(0)521-106 6888; Fax: +49-(0)521-106 2981
E-mail: joachim.heberle@uni-bielefeld.de

http://dx.doi.org/10.1016/j.jbiotec.2009.01.018

RESULTS

Journal of Biotechnology

journal homepage: www.elsevier.com/locate/jbiotec

Immobilization of the [FeFe]-hydrogenase CrHydA1 on a gold electrode: Design of a catalytic surface for the production of molecular hydrogen

Henning Krassen[a,1], Sven Stripp[b,1], Gregory von Abendroth[b], Kenichi Ataka[a], Thomas Happe[b], Joachim Heberle[a,*]

[a] Bielefeld University, Department of Chemistry, D-33615 Bielefeld, Germany
[b] Ruhr University Bochum, Department of Biochemistry of Plants, SG Photobiotechnology, D-44801 Bochum, Germany

ARTICLE INFO

Article history:
Received 14 November 2008
Received in revised form 21 January 2009
Accepted 27 January 2009

Keywords:
[FeFe]-hydrogenase
Biological hydrogen production
Monolayer
Enzyme-modified electrode
Electron transfer
FTIR
SEIRA

ABSTRACT

Hydrogenase-modified electrodes are a promising catalytic surface for the electrolysis of water with an overpotential close to zero. The [FeFe]-hydrogenase CrHydA1 from the photosynthetic green alga Chlamydomonas reinhardtii is the smallest [FeFe]-hydrogenase known and exhibits an extraordinary high hydrogen evolution activity. For the first time, we immobilized CrHydA1 on a gold surface which was modified by different carboxy-terminated self-assembled monolayers. The immobilization was in situ monitored by surface-enhanced infrared spectroscopy. In the presence of the electron mediator methyl viologen the electron transfer from the electrode to the hydrogenase was detected by cyclic voltammetry. The hydrogen evolution potential (−290 mV vs NHE, pH 6.8) of this protein modified electrode is close to the value for bare platinum (−270 mV vs NHE).
The surface coverage by CrHydA1 was determined to 2.25 ng mm^{-2} by surface plasmon resonance, which is consistent with the formation of a protein monolayer. Hydrogen evolution was quantified by gas chromatography and the specific hydrogen evolution activity of surface-bound CrHydA1 was calculated to 1.3 μmol H$_2$ min^{-1} mg^{-1} (or 85 mol H$_2$ min^{-1} mol^{-1}). In conclusion, a viable hydrogen-evolving surface was developed that may be employed in combination with immobilized photosystems to provide a platform for hydrogen production from water and solar energy with enzymes as catalysts.

© 2009 Elsevier B.V. All rights reserved.

1. Introduction

Hydrogen is one of the most promising energy carriers. Molecular hydrogen is consumed in fuel cells to generate electricity with unrivalled conversion efficiency (Cammack et al., 2001; Karyakin et al., 2005). Nowadays, the major amount of hydrogen is produced by steam methane reforming (Turner, 2004). The electrolysis of water is a costly alternative which allows the incorporation of electricity gained by renewable sources like wind and solar energy (Levene et al., 2007). For this process, rare metals, such as platinum, rhodium, or ruthenium are commonly used as catalysts because of their low overpotentials among the metallic electrocatalysts (Transatti, 1972). The overpotential on other surfaces like Ag, Cu, and Hg can be reduced by the addition of small organic molecules, like 4,4'-bipyridine (Uchida et al., 2008), thiourea (Dayalan and Narayan, 1984; Tian et al., 1990; Bukowska and Jackowska, 1994), pyridine (Hamelin et al., 1990, 1991), or methyl viologen (MV) (Tamamushi and Tanaka, 1987). Still the achieved overpotential is significantly higher than for a platinum electrode.

Nature has developed highly efficient enzymes to catalyze the reduction of protons to molecular hydrogen. These hydrogenases are able to produce hydrogen with an overpotential close (or equal) to zero (Vignais and Billoud, 2007; Armstrong and Fontecilla-Camps, 2008). Most of these enzymes are iron–sulfur proteins, which contain two metal atoms at their active site, either two iron atoms (as in [FeFe]-hydrogenases) (Peters et al., 1998; Nicolet et al., 1999) or one iron and one nickel (as in [NiFe]-hydrogenases) (Volbeda et al., 1995; Higuchi et al., 1997). [FeFe]-hydrogenases are especially interesting for biotechnological applications because of their high specific activity (Kamp et al., 2008). They are widely distributed among anaerobic living bacteria and some eukaryotic unicellular organisms (Horner et al., 2002).

In particular, hydrogen evolution of photosynthetic green algae has been an active field of basic and applied research during the last 10 years (Melis and Happe, 2001, 2004; Hemschemeier and Happe, 2005). Recent isolations of the photosynthetic hydrogenase genes (Florin et al., 2001; Winkler et al., 2002) and the discovery of H$_2$-production under sulfur deprivation in the alga Chlamydomonas reinhardtii (Zhang et al., 2001; Happe et al., 2002) gave an important impulse and opened new ways for biotechnological applications for producing hydrogen.

One approach is to fix the biological catalyst hydrogenase to an appropriate electrode. The modification of an electrode by hydro-

---

* Corresponding author. Tel.: +49 521 106 6888; fax: +49 521 106 2981.
E-mail address: joachim.heberle@uni-bielefeld.de (J. Heberle).
[1] These authors contributed equally to this work.

0168-1656/$ – see front matter © 2009 Elsevier B.V. All rights reserved.
doi:10.1016/j.jbiotec.2009.01.018

RESULTS

genase reduces the overpotential to a value close to zero and allows hydrogen production at a higher potential than on a platinum surface. This has been shown for a variety of hydrogenase enzymes including the [NiFe]-hydrogenases of *Ralstonia* species on a rotating disk graphite electrode (Goldet et al., 2008) and the [FeFe]-hydrogenases of *Desulfovibrio desulfuricans* (Vincent et al., 2005) and *Clostridium acetobutylicum* (Baffert et al., 2008). As an advantage, [FeFe]-hydrogenases have a higher hydrogen production activity (Adams, 1990; Peters et al., 1998; Nicolet et al., 1999) compared to [NiFe]-hydrogenases and suffer less of product inhibition (Léger et al., 2004). As a disadvantage, [FeFe]-hydrogenases are irreversibly inactivated by oxygen, while [NiFe]-hydrogenases can be reactivated by reduction with hydrogen or dithionite (Cammack et al., 2001).

The [FeFe]-hydrogenase of the green alga *C. reinhardtii* (HydA1) catalyses H_2-evolution with reduced methyl viologen as electron donor with a high specific activity that is comparable to the more complex [FeFe]-hydrogenases of prokaryotes (Happe and Naber, 1993). The monomeric protein of about 48 kDa only consists of the catalytically active H-cluster, but lacks any kind of accessory [FeS]-cluster (Happe and Kaminski, 2002). HydA1 is stable at ambient temperatures and deals with a wide range of buffers and salt concentrations (Forestier et al., 2003; Girbal et al., 2005). Physiologically, the hydrogenase is coupled to the photosynthetic electron transport chain via its natural electron donor, the ferredoxin PetF (Fouchard et al., 2005; Hemschemeier et al., 2008).

In this work, we present the development of an enzyme electrode which is able to catalyze the reduction of protons to molecular hydrogen at minuscule overpotential. The surface modification and immobilization of the [FeFe]-hydrogenase from *C. reinhardtii* are probed *in situ* by surface-enhanced infrared absorption spectroscopy (SEIRAS). Electrochemistry provides evidence for the catalytic activity of surface-bound CrHydA1 in the production of hydrogen. Surface plasmon resonance (SPR) and gas chromatography are used to determine the specific hydrogen evolution activity of the immobilized enzymes.

2. Material and methods

2.1. Purification of Chlamydomonas [FeFe]-hydrogenase

Recombinant [FeFe]-hydrogenase CrHydA1 Strep-tag^{exp} was produced as described before (von Abendroth et al., 2008). Briefly, *C. acetobutylicum* ATCC 824 recombinant strains were grown in CGM media in a 2.5 L MiniFors®-bioreactor (Infors, Augsburg, Germany) (Girbal et al., 2005; Wiesenborn et al., 1988). An optimized purification protocol was established for the heterologously synthesized CrHydA1 Strep-tag^{exp} enzyme. Ultracentrifugation and affinity chromatography on a 10 mL Strep-Tactin Superflow® column (IBA, Göttingen, Germany) were applied. Cell growth and protein purification were carried out under strict anaerobic conditions. Isolated protein was concentrated to 5 mg mL^{-1} on Vivaspin 6®-columns (Sartorius Stedim Biotech, Göttingen, Germany) and stored in 10% glycerol and 2 mM sodium dithionite for stabilisation. Prior to use in spectroscopic or electrochemical experiments the sample was dialysed for 30 min on 0.025 μm V-series® membranes (Millipore, Schwalbach, Germany) against 10 mM sodium phosphate buffer solution (pH 6.8).

2.2. Surface modification—monitored by surface-enhanced infrared absorption spectroscopy (SEIRAS)

A thin nano-structured gold film was chemically deposited on one side of a triangular silicon prism, as described before (Ataka and Heberle, 2003, 2007). The prism was mounted at the bottom of the spectroscopic glass cell. The infrared beam, provided from the interferometer of the FT-IR spectrometer (Bruker IFS 66 v/s, Bruker Optics, Ettlingen, Germany) was coupled into the single reflection silicon prism at an incident angle of 60°. The beam was totally internally reflected and the intensity was measured by a mercury cadmium telluride (MCT) detector (Osawa et al., 1993; Ataka et al., 1996; Ataka and Osawa, 1998, 1999; Osawa, 2002).

The bare gold surface was incubated in an aqueous solution of 2 mM mercaptopropionic acid (MPA) or an ethanolic solution of 2 mM mercaptoundecanoic acid (MUA) for 60 min (Song et al., 1993; Sun et al., 1993; Chen et al., 2002; Xu and Bowden, 2006; Jiang et al., 2008). A reference IR spectrum of the solvent-covered surface was subtracted from a series of sample spectra, which were recorded after addition of the sample. 789–1578 scans were averaged for each sample and reference spectrum, respectively. After the self-assembled monolayer (SAM) of the heterobifunctional molecules was formed, the surface was first rinsed with the solvent several times and afterwards with 10 mM sodium phosphate buffer solution (pH 6.8). The kinetics of immobilization of 170 μg mL^{-1} CrHydA1 were monitored by SEIRAS in time intervals of 30 and 60 s.

2.3. Cyclic voltammetry

A gold film or a massive gold electrode was used as a working electrode. Ag/AgCl/3M KCl and platinum mesh were used as reference and counter electrode, respectively. The cyclic voltammograms were recorded at a sweep rate of 10 mV s^{-1} with a potentiostat (Autolab PGSTAT 12, Eco Chemie B.V., Utrecht, Netherlands). All potentials are reported versus the normal hydrogen electrode (NHE).

The electrochemical experiments were performed at 20 °C in an anaerobic chamber (Coy Laboratory Products, Grassland, MI, USA) containing 95% nitrogen and 5% hydrogen. Palladium catalysts were used to remove oxygen contaminations by reduction with hydrogen. All solutions were degassed in vacuum for at least 30 min and stored in the anaerobic chamber for at least 2 weeks prior to use. Other equipment, which was transferred to the anaerobic chamber, was evacuated for at least 30 min. In this environment, the recorded cyclic voltammograms were totally stable and not influenced by oxygen during the measurement.

2.4. Amperometric hydrogen production

The electrochemical setup was embedded in a home-made, gastight measuring cell with a total volume of 20 mL. After surface modification the setup was purged with Argon for at least 10 min to remove the atmospheric hydrogen of the anaerobic chamber. 1 mL of the gas phase in the measuring cell was injected into a gas chromatograph (GC-2010, Shimadzu, Kyoto, Japan) equipped with a PLOT fused silica coating molsieve column (5 Å, 10 m by 0.32 mm; Varian, Palo Alto, CA) to gauge if the hydrogen-containing atmosphere was completely exchanged. Then a potential of −450 mV (vs NHE) was applied for 20 min while the current was monitored. 1 mL of the gas mixture was injected into a gas chromatograph to determine the amount of evolved hydrogen.

2.5. Surface plasmon resonance (SPR)

Surface plasmon resonance experiments were performed on a Biacore 3000 (GE Healthcare, Uppsala, Sweden) with a constant flow rate of 5 μL min^{-1}. Au Sensorchips were used to provide an untreated gold surface for each experiment. To form the SAM on the untreated gold surface, 300 μL of an aqueous 2 mM MPA solution were injected. For the immobilization of hydrogenase, the running buffer was changed from Millipore water to 10 mM

RESULTS

potassium phosphate buffer (pH 6.8) and 300 μL of CrHydA1 with a final concentration of 170 μg mL^{-1} were injected.

2.6. In vitro activity essay

To probe hydrogen evolution activity under optimal conditions, 1–10 μg CrHydA1 were added to 2 mL of a 100 mM sodium phosphate buffer solution (pH 6.8), containing 1 mM methyl viologen and 100 mM sodium dithionite. This solution was sealed gas-tight in an 8 mL SUBA tube, purged with argon and incubated at 37 °C for 15 min afterwards. The amount of produced hydrogen was measured by gas chromatography (as described above) and the specific hydrogen-evolving activity of the hydrogenase (in μmol H$_2$ min^{-1} mg hydrogenase^{-1}) were calculated.

3. Results and discussion

3.1. Immobilization of the hydrogenase

To design a surface which is able to reduce protons to molecular hydrogen, the catalyst, the [FeFe]-hydrogenase from *C. reinhardtii* CrHydA1 was immobilized on the surface of a solid gold electrode by electrostatic interaction. Immobilization was followed *in situ* by surface-enhanced IR absorption spectroscopy. Fig. 1 depicts the SEIRA spectra of CrHydA1 during the adsorption to the mercaptopropionic acid-modified surface (A). In Fig. 1(A) bands arise at 1659 and 1550 cm^{-1}. These bands are assigned to the amide I and amide II modes of the protein backbone (Krimm and Bandekar, 1986). The band intensities reflect the amount of protein adsorbed to the surface and, therefore, increase during the adsorption process. While the proteins bind to the surface, a negative band arises at frequencies > 1700 cm^{-1}, which overlaps with the amide I band. This band is assigned to water (H–O–H bending mode), which is displaced from the vicinity of the surface. Signals from the bulk phase are negligible as the surface-enhancement decays exponentially with distance (decay length ∼10 nm). The absence of any amide bands

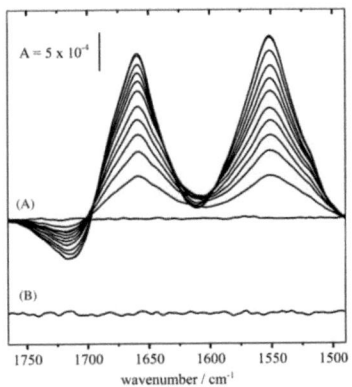

Fig. 1. (A) SEIRA spectra of the binding of CrHydA1 to a MPA-modified gold electrode. The displayed spectra are recorded at 0, 0.5, 1, 2, 5, 10, 20, 30, 45, 60, 75, and 90 min after addition of the protein. The rising bands indicate the binding process and are discussed in the text. (B) SEIRA spectrum of a bare gold surface after incubation with CrHydA1 for 30 min. The absence of bands indicates that CrHydA1 does not bind to the bare gold surface.

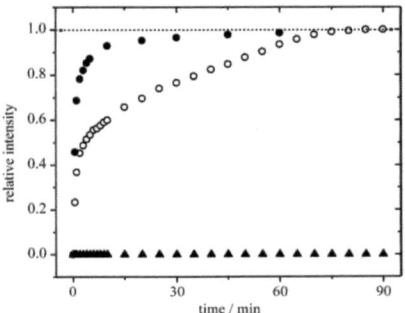

Fig. 2. Adsorption kinetics of 170 μg mL^{-1} CrHydA1 to an MUA-SAM (●), an MPA-SAM (○) and a bare gold surface (▲) are compared. The peak height of the amide II band at 1550 cm^{-1} is normalized to 1 for maximum coverage and plotted versus the adsorption time.

upon injecting a similar protein solution to an unmodified gold surface in Fig. 1(B) shows that CrHydA1 does not bind to a bare gold surface.

The gold surface was modified by heterobifunctional molecules to increase the affinity of CrHydA1 to the surface. Mercaptopropionic acid or mercaptoundecanoic acid was used to form a carboxy-terminated SAM. At the pH of 6.8, the surface is negatively charged and allows electrostatic binding of CrHydA1. The strength of the electrostatic interaction varies with the chosen SAM.

The adsorption kinetics of CrHydA1 to different surfaces are shown in Fig. 2. The peak height of the amide II band is normalized to 1 for maximum coverage and plotted versus the adsorption time. 90% of the maximum coverage with CrHydA1 on an MUA-SAM is reached after 5 min, while it takes 50 min on a MPA-SAM, which is also terminated by carboxy-groups, but less flexible due to the shorter chain length.

In both cases, the immobilized protein film is stable. Rinsing with buffer (10 mM potassium phosphate, pH 6.8) only removes the unspecifically bound protein. The peak height of amides I and II band decreases by less than 5% at the beginning, while further washing does not lead to further decrease in intensity (data not shown).

3.2. Probing the catalytic activity of the surface-bound CrHydA1

The electrochemical activity of CrHydA1-modified gold surfaces is compared among the different adsorption conditions by means of cyclic voltammetry. In the cyclic voltammograms of CrHydA1 on an MUA-modified surface (Fig. 3(A), dotted curve), no increase in the reductive current is observed. This indicates that electrons are not directly transferred from the electrode to the hydrogenase. After addition of the electron mediator methyl viologen the reductive current increases (solid curve) due to electron transfer from the electrode to MV^{2+}. When increasing the potential, no oxidation peak appears, indicating that MV is oxidized by transferring electrons to the hydrogenase, where electrons are used to reduce protons to molecular hydrogen.

Fig. 3(B) shows the cyclic voltammogram of CrHydA1 on a bare gold electrode (dotted curve). If electrons would be transferred to the hydrogenase the amplitude of the reductive ("negative") current would increase below the necessary potential. The absence of this feature proves that there is no direct electron transfer from a

23

RESULTS

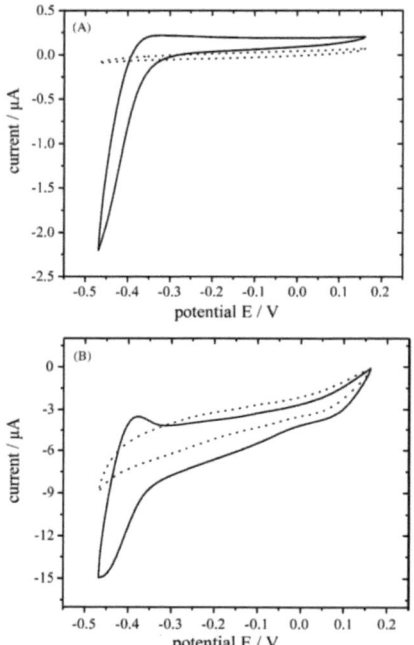

Fig. 3. (A) CrHydA1 is immobilized on an MUA–SAM. The dotted curve shows the cyclic voltammogram before addition of 100 μM MV, the solid curve after the addition. (B) Cyclic voltammograms of CrHydA1 on a bare gold electrode. The dotted line is measured before, the solid line 5 min after addition of 100 μM MV.

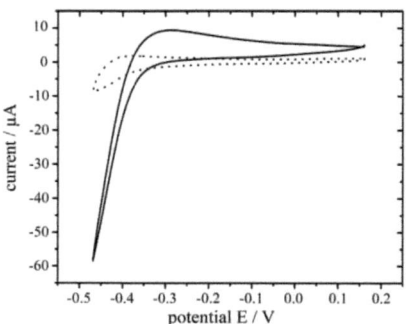

Fig. 4. Cyclic voltammogram of a 100 μM MV solution on a MPA-modified gold electrode (dotted curve) compared to the cyclic voltammogram 10 min after addition of 170 μg mL^{-1} CrHydA1 (solid curve).

bare gold electrode to CrHydA1 in solution. The addition of MV^{2+} (solid curve) leads to a pair of reduction and oxidation peak with a mid-point potential of −423 mV, as expected from literature for MV (Stombaugh et al., 1976). The oxidation peak at −383 mV is assigned to the re-oxidation of MV at the electrode and its appearance shows, that MV is not (or at a negligible rate) oxidized by the hydrogenase. No (or negligible) mediated electron transfer takes place between a bare gold electrode and CrHydA1, if the protein is not immobilized on the surface.

An increase of the reductive current after addition of CrHydA1 (Fig. 4, solid curve) to a 100 μM solution of MV^{2+} on an MPA-modified electrode (dotted curve) is observed. Current in the dotted curve is comparable to the cyclic voltammogram of MV as displayed in Fig. 3(B). The reductive current at a potential of −450 mV reaches a value of −47 μA, while it is only −2 μA on a MUA-modified electrode (Fig. 3(A)).

Two factors explain the difference in the catalytic activity of the immobilized hydrogenase on different SAMs: (1) Due to different chain length, the MUA–SAM is more flexible than the MPA–SAM. This results in a different surface structure and probably a different orientation of the hydrogenase. If the electron acceptor site of CrHydA1 gets closer to the surface of the SAM, the accessibility of the acceptor site might be restricted by sterical hindrance. This would result in a smaller reductive current. (2) MV^{2+} is not reduced on top of the SAM but partially penetrates the monolayer (see Supplementary informations). The diffusion distance from the place where MV^{2+} is reduced to the electron acceptor site of CrHydA1 is larger for the MUA–SAM. This would result in a smaller reductive current as well.

3.3. Quantification of the hydrogen production

Cyclic voltammetry showed that electrons are transferred to the hydrogenase. However, it is not clear if the transferred electrons are consumed to reduce protons and produce hydrogen. This can be demonstrated by the direct detection of molecular hydrogen by gas chromatography. CrHydA1 is immobilized on a MPA-modified gold film electrode and covered with the MV-containing potassium phosphate buffer (pH 6.8). After a constant potential of −450 mV has been applied to this system for 20 min, 1 mL of the gas phase is injected into the gas chromatograph. The area of the hydrogen peak is calculated by integration to an average area of 140, which equates 20 nmol H$_2$ (n(H$_2$)) in the total volume.

While the potential is applied, the current is monitored by amperometry. The transferred charge is calculated to 25 mC by integration over 20 min, which equals a maximum theoretical H$_2$-production of n_{max}(H$_2$) = 130 nmol, if all electrons are transferred to the hydrogenase. The catalytic efficiency $\eta_{cat} = n$(H$_2$)/n_{max}(H$_2$) is calculated to 15% at the applied potential of −450 mV.

Two factors contribute to the difference between calculated and measured hydrogen: (1) Electrons can be conducted from the gold electrode to the counter electrode without being used by the hydrogenase. (2) A fraction of the surface might be covered with protein which does not produce molecular hydrogen – either denatured during purification/dialysis or bound in an orientation which blocks the electron acceptor site – and transfers the electrons to other acceptors in the solution.

Just recently, Hambourger et al. reported immobilization of the [FeFe]-hydrogenase CaHydA from C. acetobutylicum on glassy carbon and carbon felt (Hambourger et al., 2008). Proteins can bind directly on the surface and receive electrons via direct electron transfer on certain forms of graphite. However, quantification of the bound enzymes is not possible on a graphite electrode, but on a gold electrode as presented here.

RESULTS

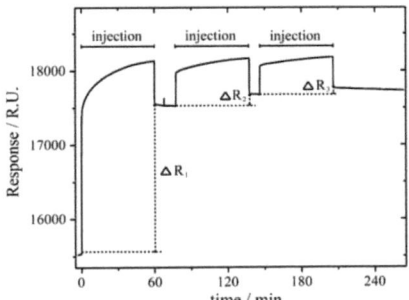

Fig. 5. Surface plasmon resonance signal of the binding of CrHydA1 on a MPA-modified gold surface. The surface is continuously rinsed with buffer at a flow rate of 5 μL min^{-1}. At $t=0$, 77, and 146 min CrHydA1 is injected for 1 h, respectively. The increase in the SPR response is indicated as ΔR_n for each injection. ($\Delta R_1 = 2014$ R.U., $\Delta R_2 = 162$ R.U., $\Delta R_3 = 78$ R.U.).

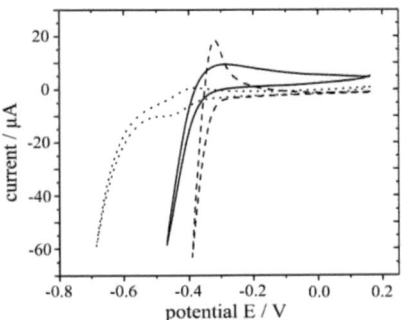

Fig. 6. Cyclic voltammograms of hydrogen evolution in sodium phosphate buffer (pH 6.8) on the bare platinum electrode (dashed curve), on the bare gold electrode with MV in solution (dotted curve), and on the CrHydA1-modified electrode (solid curve). In the latter case, methyl viologen is used as soluble electron carrier. The measured current for the platinum electrode is divided by 20 for better comparison.

3.4. Specific activity of immobilized hydrogenase

The specific hydrogen-evolving activity of recombinant CrHydA1 in solution is measured by established *in vitro* tests to 760 μmol H$_2$ min^{-1} mg^{-1} (Girbal et al., 2005). After dialysis, the average specific activity of the CrHydA1 samples used in our experiments drops to 130 μmol H$_2$ min^{-1} mg^{-1} corresponding to 8500 mol H$_2$ min^{-1} mol^{-1}. This value defines the upper limit of the specific hydrogen-evolving activity of the immobilized CrHydA1.

Surface plasmon resonance was used to quantify the amount of immobilized CrHydA1. Binding of the protein to the MPA-modified surface and the exchange of the buffer with the protein solution change the refractive index and lead to an increase of the SPR response, respectively (Fig. 5). Before and after the injection, the surface is rinsed with buffer and the difference in the SPR response is solely attributed to the bound protein. These values are noted in Fig. 5 as ΔR_n for each protein injection. During the first injection the entire surface of the MPA-SAM is available, while in the following injections only the uncovered parts can bind proteins. Consequently, ΔR_2 and ΔR_3 are much smaller than ΔR_1. The sum of all three $\Delta R_n \approx 2250$ R.U. reflects the total amount of specifically bound hydrogenase and is calculated to 2.25 ng CrHydA1 mm^{-2} (or 3.42×10^{-12} mol cm^{-2}) with the conversion of 1000 R.U. into 1 ng protein mm^{-2} (Armstrong et al., 1947; Stenberg et al., 1991). The measured value indicates that a protein monolayer is adsorbed to the surface and the contribution of unspecific multilayer is negligible.

In the amperometric experiment 20 nmol H$_2$ have been produced in 20 min. The electrode used has a geometrical surface area of 1.45 cm^2 and a surface roughness of 2.5 (Miyake et al., 2002). From these values, the specific hydrogen-evolving activity of surface-bound CrHydA1 can be calculated to 1.3 μmol H$_2$ min^{-1} mg^{-1} or 85 mol H$_2$ min^{-1} mol^{-1}, which is 1% of the activity in the *in vitro* test.

For the *in vitro* essay, the strong reductant sodium dithionite is added in excess to immediately re-reduce oxidized MV^{2+} in the vicinity of the electron acceptor site of CrHydA1 and maintains a constant high concentration of reduced MV. Using a protein monolayer, re-reduction of MV^{2+} takes place close to the electrode surface (tunnelling distance) and the reduced MV diffuses a longer distance to the hydrogenase, which limits the reaction rate. Another explanation for the comparatively low activity of the immobilized CrHydA1 is that the access to the electron acceptor site of CrHydA1 might be hindered by the MPA-SAM, compared to the protein in solution (see above). In addition the temperature is 20 °C for the monolayer experiments and 37 °C for the *in vitro* tests. This results in a reaction rate lowered by a factor of 3 (from thermodynamics).

3.5. Hydrogen evolution potential

The hydrogen evolution potential is defined as the most positive potential, which allows hydrogen production at the given surface. In Fig. 6 three cyclic voltammograms (pH 6.8) are compared on different electrodes. Each exhibits an increase in the reductive current, at potentials below its hydrogen evolution potential. On a bare gold electrode with methyl viologen as electron mediator, hydrogen evolution takes place below −460 mV (dotted line in Fig. 6). By modifying the gold surface with hydrogenase like CrHydA1, the hydrogen evolution potential is improved to −290 mV (solid line). This characteristic value is close to the potential of −270 mV on platinum surfaces (dashed line), which are used in the industry for electrochemical hydrogen production.

In summary, we demonstrated the successful immobilization of the [FeFe]-hydrogenase CrHydA1 from *C. reinhardtii* on a carboxy-terminated self-assembled monolayer. For the first time, a "wireless" [FeFe]-hydrogenase (i.e. without electron relay chain) was examined by protein film voltammetry. If methyl viologen is used as a soluble electron carrier, the CrHydA1-modified surface was able to catalyze the reduction of protons to molecular hydrogen at a similar potential as on platinum electrodes. At a potential of −450 mV 15–17% of the provided electrons are used for hydrogen production.

The possibility to investigate the bound hydrogenase by surface-enhanced infrared spectroscopy and to quantify the amount of bound protein via SPR makes the system a promising approach for in-depth analysis of the specific activity per molecule. This will make it possible to distinguish between different surface populations of the hydrogenase and optimize the electrode to highest activity. The system can be used to probe the specific redox activity and efficiency of variable hydrogenase and protein films in general – as a platform technology for further investigations.

RESULTS

Acknowledgements

We gratefully acknowledge financial support by the BMBF (Grundlagen für einen biotechnologischen und biomimetischen Ansatz der Wasserstoffproduktion). S. Stripp and T. Happe were further supported by the Deutsche Forschungsgemeinschaft (SFB 480). We thank Prof. Dr. N. Sewald and K. Wollschläger for introduction to and use of the Biacore 3000 and discussions.

Appendix A. Supplementary data

Supplementary data associated with this article can be found, in the online version, at doi:10.1016/j.jbiotec.2009.01.018.

References

Adams, M.W., 1990. The structure and mechanism of iron-hydrogenases. Biochim. Biophys. Acta 1020 (2), 115–145.
Armstrong, F.A., Fontecilla-Camps, J.C., 2008. Biochemistry. A natural choice for activating hydrogen. Science 321, 498–499.
Armstrong, S.H., Budka, M.J.E., Morrison, K.C., Hasson, M., 1947. Preparation and properties of serum and plasma proteins. XII. The refractive properties of the proteins of human plasma and certain purified fractions. J. Am. Chem. Soc. 69, 1747–1753.
Ataka, K., Yotsuyanagi, T., Osawa, M., 1996. Potential-dependent reorientation of water molecules at an electrode/electrolyte interface studied by surface-enhanced infrared absorption spectroscopy. J. Phys. Chem. 100, 10664–10672.
Ataka, K., Osawa, M., 1998. In situ infrared study of water-sulfate coadsorption on gold(111) in sulfuric acid solutions. Langmuir 14 (4), 951–959.
Ataka, K., Osawa, M., 1999. In situ infrared study of cytosine adsorption on gold electrodes. J. Electroanal. Chem. 460, 188–196.
Ataka, K., Heberle, J., 2003. Electrochemically induced surface-enhanced infrared difference spectroscopy (SEIDA) spectroscopy of a protein monolayer. J. Am. Chem. Soc. 125, 4986–4987.
Ataka, K., Heberle, J., 2007. Biochemical applications of surface-enhanced infrared absorption spectroscopy. Anal. Bioanal. Chem. 388 (1), 47–54.
Baffert, C., Demuez, M., Cournac, L., Burlat, B., Guigliarelli, B., Bertrand, P., Girbal, L., Léger, C., 2008. Hydrogen-activating enzymes: activity does not correlate with oxygen sensitivity. Angew. Chem. 120 (11), 2082–2084.
Bukowska, J., Jackowska, K., 1994. Influence of thiourea on hydrogen evolution at a silver electrode as studied by electrochemical and SERS methods. J. Electroanal. Chem. 367, 41–48.
Cammack, R., Frey, M., Robson, R., 2001. Producing hydrogen as a fuel. In: Cammack, R., Frey, M., Robson, R. (Eds.), Hydrogen as a Fuel: Learning from Nature. Routledge, UK, pp. 201–230.
Chen, X., Ferrigno, R., Yang, J., Whitesides, G.A., 2002. Redox properties of cytochrome c adsorbed on self-assembled monolayers: a probe for protein conformation and orientation. Langmuir 18 (18), 7009–7015.
Dayalan, E., Narayan, R., 1984. The acceleration of the hydrogen evolution reaction from acid solutions by sulfur compounds. Part I. Mercury electrode. J. Electroanal. Chem. 179, 167–178.
Florin, T., Tsokoglou, A., Happe, T., 2001. A novel type of iron hydrogenase in the green alga Scenedesmus obliquus is linked to the photosynthetic electron transport chain. J. Biol. Chem. 276, 6125–6132.
Forestier, M., King, P., Zhang, L., Posewitz, M., Schwarzer, S., Happe, T., Ghirardi, M.L., Seibert, M., 2003. Expression of two [Fe]-hydrogenases in Chlamydomonas reinhardtii under anaerobic conditions. Eur. J. Biochem. 270 (13), 2750–2758.
Fouchard, S., Hemschemeier, A., Caruana, A., Pruvost, J., Legrand, J., Happe, T., Peltier, G., Cournac, L., 2005. Investigations on autotrophic and mixotrophic hydrogen photoproduction in sulphur-deprived Chlamydomonas cells. Appl. Environ. Microbiol. 71, 6199–6205.
Girbal, L., von Abendroth, G., Winkler, M., Benton, P.M.C., Meynial-Salles, I., Croux, C., Peters, J.W., Happe, T., Soucaille, P., 2005. Homologous and heterologous overexpression in Clostridium acetobutylicum and characterization of purified clostridial and algal Fe-only hydrogenases with high specific activities. Appl. Environ. Microbiol. 71 (5), 2777–2781.
Goldet, G., Wait, A.F., Cracknell, J.A., Vincent, K.A., Ludwig, M., Lenz, O., Friedrich, B., Armstrong, F.A., 2008. Hydrogen production under aerobic conditions by membrane-bound hydrogenases from Ralstonia species. J. Am. Chem. Soc. 130 (33), 11106–11113.
Hambourger, M., Gervaldo, M., Svedruzic, D., King, P.W., Gust, D., Ghirardi, M., Moore, A.L., Moore, T.A., 2008. [FeFe]-hydrogenase-catalyzed H_2 production in a photoelectrochemical biofuel cell. J. Am. Chem. Soc. 130 (6), 2015–2022.
Hamelin, A., Morin, S., Richer, J., Lipkowski, J., 1990. Adsorption of pyridine on the (311) face of silver. J. Electroanal. Chem. 285, 249–262.
Hamelin, A., Morin, S., Richer, J., Lipkowski, J., 1991. Adsorption of pyridine on the (210) face of silver. J. Electroanal. Chem. 304, 195–209.
Happe, T., Hemschemeier, A., Kaminski, A., 2002. Hydrogenases in green algae: do they save the algae's life and solve our energy problems? Trends Plant Sci. 7, 246–250.

Happe, T., Kaminski, A., 2002. Differential regulation of the Fe-hydrogenase during anaerobic adaptation in the green alga Chlamydomonas reinhardtii. Eur. J. Biochem. 269 (3), 1022–1032.
Happe, T., Naber, J.D., 1993. Isolation, characterization and N-terminal amino acid sequence of hydrogenase from the green alga Chlamydomonas reinhardtii. Eur. J. Biochem. 214 (2), 475–481.
Hemschemeier, A., Fouchard, S., Cournac, L., Peltier, G., Happe, T., 2008. Hydrogen production by Chlamydomonas reinhardtii: an elaborate interplay of electron sources and sinks. Planta 227, 397–407.
Hemschemeier, A., Happe, T., 2005. The exceptional photofermentative hydrogen metabolism of the green alga Chlamydomonas reinhardtii. Biochem. Soc. Trans. 33, 39–41.
Higuchi, Y., Yagi, T., Yasuoka, N., 1997. Unusual ligand structure in Ni–Fe active center and an additional Mg site in hydrogenase revealed by high resolution X-ray structure analysis. Structure 5, 1671–1680.
Horner, D., Heil, B., Happe, T., Embley, M., 2002. Iron hydrogenases, ancient enzymes in modern eukaryotes. Trends Biochem. Sci. 27, 148–153.
Jiang, X., Ataka, K., Heberle, J., 2008. Influence of the molecular structure of carboxyl-terminated self-assembled monolayer on the electron transfer of cytochrome c adsorbed on an Au electrode: In situ observation by surface-enhanced infrared absorption spectroscopy. J. Phys. Chem. C 112 (3), 813–819.
Kamp, C., Silakov, A., Winkler, M., Reijerse, E.J., Lubitz, W., Happe, T., 2008. Isolation and first EPR characterization of the [FeFe]-hydrogenases from green algae. Biochim. Biophys. Acta 1777, 410–416.
Karyakin, A.A., Morozov, S.V., Karyakina, E.E., Zorin, N.A., Perelygin, V.V., Cosnier, S., 2005. Hydrogenase electrodes for fuel cells. Biochem. Soc. Trans. 33, 73–75.
Krimm, S., Bandekar, J., 1986. Vibrational spectroscopy and conformation of peptides, polypeptides, and proteins. Adv. Protein Chem. 38, 181–364.
Léger, C., Dementin, S., Bertrand, P., Rousset, M., Guigliarelli, B., 2004. Inhibition and aerobic inactivation kinetics of Desulfovibrio fructosovorans NiFe hydrogenase studied by protein film voltammetry. J. Am. Chem. Soc. 126, 12162–12172.
Levene, J.I., Manna, M.K., Margolisa, R.M., Milbrandt, A., 2007. An analysis of hydrogen production from renewable electricity sources. Sol. Energ. 81 (6), 773–780.
Melis, A., Happe, T., 2001. Hydrogen production. Green algae as a source of energy. Plant Physiol. 179, 740–748.
Melis, A., Happe, T., 2004. Trails of green alga H_2-production research – from Hans Gaffron to new frontiers. Photosyn. Res. 80, 401–409.
Miyake, H., Ye, S., Osawa, M., 2002. Electroless deposition of gold thin films on silicon for surface-enhanced infrared spectroelectrochemistry. Electrochem. Commun. 4 (12), 973–977.
Nicolet, Y., Piras, C., Legrand, P., Hatchikian, C.E., Fontecilla-Camps, J.C., 1999. Desulfovibrio desulfuricans iron hydrogenase: the structure shows unusual coordination to an active site Fe binuclear center. Structure 7, 13–23.
Osawa, M., Ataka, K., Yoshii, K., Yotsuyanagi, T.J., 1993. Surface-enhanced infrared ATR spectroscopy for in situ studies of electrode/electrolyte interfaces. Electron Spectrosc. Relat. Phenom. 64–65, 371–379.
Osawa, M., 2002. Surface-enhanced infrared absorption spectroscopy. In: Chalmers, J.M., Griffiths, P.R. (Eds.), Handbook of Vibrational Spectroscopy. Wiley, Chichester, U.K., pp. 785–799.
Peters, J.W., Lanzilotta, W.N., Lemon, B.J., Seefeldt, L.C., 1998. X-ray crystal structure of the Fe-only hydrogenase (CpI) from Clostridium pasteurianum to 1.8 angstrom resolution. Science 282, 1853–1858.
Song, S., Clark, R.A., Bowden, E.F., Tarlov, M.J., 1993. Characterization of cytochrome c/alkanethiolate structures prepared by self-assembly on gold. J. Phys. Chem. 97 (24), 6564–6572.
Stenberg, E., Persson, B., Roos, H., Urbaniczky, C., 1991. Quantitative determination of surface concentration of protein with surface plasmon resonance using radiolabeled proteins. J. Colloid Interface Sci. 143, 513–526.
Stombaugh, N.A., Sundquist, J.E., Burris, R.H., Orme-Johnson, W.H., 1976. Oxidation-reduction properties of several low potential iron-sulfur proteins and of methylviologen. Biochemistry 15 (12), 2633–2641.
Sun, L., Crooks, R.M., Ricco, A.J., 1993. Molecular interactions between organized, surface-confined monolayers and vapor-phase probe molecules. 5. Acid–base interactions. Langmuir 9 (7), 1775–1780.
Tamamushi, R., Tanaka, K., 1987. Effect of methylviologen on the hydrogen evolution at mercury electrodes in aqueous buffer solutions. J. Electroanal. Chem. 230, 177–188.
Tian, Z.Q., Lian, Y.Z., Fleischmann, M., 1990. In-situ raman spectroscopic studies on coadsorption of thiourea with anions at silver electrodes. Electrochim. Acta 35 (5), 879–883.
Transatti, S., 1972. Work function, electronegativity, and electrochemical behaviour of metals. III. Electrolytic hydrogen evolution in acid solutions. J. Electroanal. Chem. 39 (1), 163–184.
Turner, J.A., 2004. Sustainable hydrogen production. Science 305, 972–974.
Uchida, T., Mogami, H., Yamakata, A., Sasaki, Y., Osawa, M., 2008. Hydrogen evolution reaction catalyzed by proton-coupled redox cycle of 4,4′-bipyridine monolayer adsorbed on silver electrodes. J. Am. Chem. Soc. 130, 10862–10863.
Vignais, P.M., Billoud, B., 2007. Occurrence, classification, and biological function of hydrogenases: an overview. Chem. Rev. 107, 4206–4272.
Vincent, K.A., Parkin, A., Lenz, O., Albracht, S.P.J., Fontecilla-Camps, J.C., Cammack, R., Friedrich, B., Armstrong, F.A., 2005. Electrochemical definitions of O_2 sensitivity and oxidative inactivation in hydrogenases. J. Am. Chem. Soc. 127 (51), 18179–18189.

RESULTS

Volbeda, A., Charon, M.H., Piras, C., Hatchikian, C.E., Frey, M., Fontecilla-Camps, J.C., 1995. Crystal structure of the nickel-iron hydrogenase from *Desulfovibrio gigas*. Nature 373, 580-587.

von Abendroth, G., Stripp, S., Silakov, A., Croux, C., Soucaillec, P., Girbal, L., Happe, T., 2008. Optimized over-expression of [FeFe] hydrogenases with high specific activity in *Clostridium acetobutylicum*. Int. J. Hydrogen Energ. 33 (21), 6076-6081.

Wiesenborn, D.P., Rudolph, F.B., Papoutsakis, E.T., 1988. Thiolase from *Clostridium acetobutylicum* ATCC 824 and its role in the synthesis of acids and solvents. Appl. Environ. Microbiol. 54 (11), 2717-2722.

Winkler, M., Heil, B., Heil, B., Happe, T., 2002. Isolation and molecular characterization of the [Fe]-hydrogenase from the unicellular green alga *Chlorella fusca*. Biochim. Biophys. Acta 1576, 330-334.

Xu, J., Bowden, E.F., 2006. Determination of the orientation of adsorbed cytochrome c on carboxyalkanethiol self-assembled monolayers by in situ differential modification. J. Am. Chem. Soc. 128 (21), 6813-6822.

Zhang, L., Happe, T., Melis, A., 2001. Biochemical and morphological characterization of sulfur-deprived and H_2-producing *Chlamydomonas reinhardtii*. Planta 214, 552-561.

RESULTS

2.3 The Structure of the Active Site H-Cluster of [FeFe] Hydrogenase from the Green Alga *Chlamydomonas reinhardtii* Studied by X-Ray Absorption Spectroscopy

Sven T. Stripp[1§] & Oliver Sanganas[2§], Thomas Happe[1*], Michael Haumann[2*&]

[1] Ruhr-Universität Bochum, Lehrstuhl für Biochemie der Pflanzen, AG Photobiotechnologie, 44780 Bochum, Germany

[2] Freie Universität Berlin, Institut für Experimentalphysik, Arnimallee 14, 14195 Berlin, Germany

[§] These authors contributed equally to this work.

*** Corresponding authors:**
Prof. Thomas Happe, Ruhr-Universität Bochum
Phone: +49 234 32 27026; Fax: +49 234 32 14322
E-mail: thomas.happe@rub.de
Dr. Michael Haumann, Freie Universität Berlin
Phone: +49 30 838 56101; Fax: +49 30 838 56299,
E-mail: michael.haumann@fu-berlin.de

http://dx.doi.org/ 10.1021/bi900010b

RESULTS

The Structure of the Active Site H-Cluster of [FeFe] Hydrogenase from the Green Alga *Chlamydomonas reinhardtii* Studied by X-ray Absorption Spectroscopy[†]

Sven Stripp,[‡,∥] Oliver Sanganas,[§,∥] Thomas Happe,*[,‡] and Michael Haumann*[,§]

[‡]*Lehrstuhl für Biochemie der Pflanzen, AG Photobiotechnologie, Ruhr-Universität Bochum, 44780 Bochum, Germany, and* [§]*Institut für Experimentalphysik, Freie Universität Berlin, Arnimallee 14, 14195 Berlin, Germany* [∥]*These authors contributed equally to this work*

Received January 5, 2009; Revised Manuscript Received April 14, 2009

ABSTRACT: The [FeFe] hydrogenase (*Cr*HydA1) of the green alga *Chlamydomonas reinhardtii* is the smallest hydrogenase known and can be considered as a "minimal unit" for biological H_2 production. Due to the absence of additional FeS clusters as found in bacterial [FeFe] hydrogenases, it was possible to specifically study its catalytic iron–sulfur cluster (H-cluster) by X-ray absorption spectroscopy (XAS) at the Fe K-edge. The XAS analysis revealed that the *Cr*HydA1 H-cluster consists of a [4Fe4S] cluster and a diiron site, $2Fe_H$, which both are similar to their crystallographically characterized bacterial counterparts. Determination of the individual Fe–Fe distances in the [4Fe4S] cluster (~2.7 Å) and in the $2Fe_H$ unit (~2.5 Å) was achieved. Fe–C (=O/N) and Fe–S bond lengths were in good agreement with crystallographic data on bacterial enzymes. The loss of Fe–Fe distances in protein purified under mildly oxidizing conditions indicated partial degradation of the H-cluster. Bond length alterations detected after incubation of *Cr*HydA1 with CO and H_2 were related to structural and oxidation state changes at the catalytic Fe atoms, e.g., to the binding of an exogenous CO at $2Fe_H$ in CO-inhibited enzyme. Our XAS studies pave the way for the monitoring of atomic level structural changes at the H-cluster during H_2 catalysis.

Hydrogenases are efficient catalysts for the production and cleavage of molecular hydrogen (H_2) (*1*, *2*). Their use in biotechnological applications is expected to significantly contribute to future renewable fuel production (*3*, *4*). The molecular principles of H_2 turnover in these enzymes may lead to novel non-platinum catalysts (*5*). It is thus an important challenge to unravel the reaction mechanism of H_2 catalysis in hydrogenases.

In biocatalysis the [FeFe] hydrogenases are of particular interest as they show extremely high rates of H_2 production (*6*, *7*). The structures of [FeFe] hydrogenases from two bacterial species, namely, *Clostridium pasteurianum* and *Desulfovibrio desulfuricans*, have been resolved by protein crystallography (*8*, *9*). Their active site is a unique iron–sulfur cluster, comprising six Fe atoms commonly known as the H-cluster (*6*, *10*). It can be divided into a [4Fe4S] cluster and a binuclear iron unit, $2Fe_H$,[1] cysteine-linked to the cubane. This diiron moiety is probably the active site of H_2 catalysis (*8*, *11*, *12*). This situation differs from that in the [NiFe] hydrogenases where a heterobimetallic complex forms the active site (*13*). Depending on the redox conditions, the two Fe atoms of $2Fe_H$ bind three to four CO and two CN ligands. One CO may be found in an Fe–Fe bridging position (*14*–*17*). An azadithiolate (adt) moiety has been proposed, but the precise chemical identity of this ligand on top of the $2Fe_H$ site is not yet settled (*12*, *18*). Besides the H-cluster, bacterial [FeFe] hydrogenases harbor up to four additional FeS clusters, serving as electron transfer relays (*19*). Signal overlay from these clusters complicates investigations on the reaction cycle of the bacterial H-cluster by Fe-specific techniques, i.e., Mössbauer (*20*) and X-ray absorption spectroscopy (XAS) (*18*, *21*, *22*).

[FeFe] hydrogenases exist not only in anaerobic bacteria but also in photosynthetic eukaryotes. For several green algae, the *hydA* genes and respective proteins have been identified (*23*–*26*). These extraordinarily small proteins may be regarded as a "minimal unit" of biological H_2 conversion. Sequence alignments show that the small F-domain carrying the FeS relay clusters in bacteria is missing (*24*, *25*). However, four cysteine residues necessary for H-cluster accommodation in the H-domain are well conserved in all hydrogenase-coding *hydA* genes (*23*).

The biophysical characterization of algal [FeFe] hydrogenases is hampered by the fact that these enzymes are extremely O_2 sensitive and synthesized *in vivo* only in small amounts (*23*, *28*). To overcome low protein yields, we have established a general

[†]Financial support by the Deutsche Forschungsgemeinschaft (SFB498-C8, M.H.; SFB480, T.H.), the "Unicat" Cluster of Excellence Berlin, the Bundesministerium für Bildung und Forschung (BioH2 program), and the EU/Energy Network SolarH2 (FP7 contract 212508) is gratefully acknowledged.
*Corresponding authors. T.H.: phone: +49 234 32 27026; fax: +49 234 32 14322; e-mail: thomas.happe@rub.de. M.H.: phone: +49 30 838 56101; fax: +49 30 838 56299; e-mail: michael.haumann@fu-berlin.de.
[1]Abbreviations: $2Fe_H$, binuclear Fe unit of the H-cluster; adt, azadithiolate; AI, as-isolated state; CpI, *Clostridium pasteurianum* [FeFe] hydrogenase; DdH, *Desulfovibrio desulfuricans* [FeFe] hydrogenase; EPR, electron paramagnetic resonance spectroscopy; EXAFS, extended X-ray absorption fine structure; FTIR, Fourier-transform infrared spectroscopy; *Cr*HydA1, [FeFe] hydrogenase protein of *C. reinhardtii*; MS, multiple scattering; XAS, X-ray absorption spectroscopy.

system for heterologous expression and synthesis of [FeFe] hydrogenases in *Clostridium acetobutylicum*. The [FeFe] hydrogenase from *Chlamydomonas reinhardtii*, CrHydA1, is efficiently assembled due to the maturation apparatus of the host organism, yielding 2 mg of protein/L of cell culture (28). Hence, a sufficient amount of CrHydA1 [FeFe] hydrogenase is available for in-depth spectroscopic explorations.

CrHydA1 is an ideal candidate to study the assembly, structure, and catalytic mechanism of [FeFe] hydrogenases. First, EPR studies on CrHydA1 and two further algal [FeFe] hydrogenases revealed similar signals of their CO-inhibited state of the active site as found in bacterial enzymes. However, deviations in the EPR g-tensors suggested distinct differences in the electronic structure of the H-cluster (27). Direct information on the geometric structure of the CrHydA1 H-cluster is required to compare it to the situation in bacteria.

In the present study X-ray absorption spectroscopy (XAS) (29) at the Fe K-edge was performed to obtain atomic level structural and electronic information on the H-cluster of CrHydA1. We show that the structural parameters imply an overall organization of the algal H-cluster similar to that in bacterial enzymes. Structural alterations induced by the inhibitor CO and the substrate H_2 are addressed.

MATERIALS AND METHODS

Protein Sample Preparation. C. reinhardtii [FeFe] hydrogenase CrHydA1 was heterologously synthesized and isolated as previously described (30). C. acetobutylicum ATCC 824 recombinant strains were grown in CGM media containing up to 60 g/L glucose in a 2.5 L MiniFors bioreactor (Infors, Augsburg, Germany) (28, 31). Both cell growth and protein purification were carried out under strictly anoxic conditions. Isolated protein was concentrated up to 1 mM (about 48 mg/mL) on Vivaspin 6 and Vivaspin 500 columns (Sartorius Stedim Biotech, Göttingen, Germany) and stored in 10% glycerol and 2 mM sodium dithionite (NaDT) for stabilization.

The specific H_2-evolving activity was determined by an *in vitro* assay as described before (23). The gas mixture was injected into a gas chromatograph (GC-2010; Shimadzu, Kyoto, Japan) equipped with a PLOT fused silica coating Molsieve column (5 Å, 10 m by 0.32 mm) from Varian (Palo Alto, CA). The specific activity of the hydrogenase was calculated from the detected amount of produced H_2.

For gas treatments, protein samples (~1 mM) were filled into open 200 μL PCR tubes and placed in 8 mL Suba tubes, which were rubber sealed under the N_2/H_2 atmosphere of the anaerobic tent. To humidify the working gases (CO, H_2), degassed water was bubbled with each of the two gases for 30 min, respectively. The headspace of the hermetically sealed Suba tubes was flushed with gas for 15–30 min. The final protein concentration of the samples used at the DESY Hamburg was adjusted right after the gas treatments; hence, more diluted protein was flushed. During the treatments, protein was kept on ice and protected against light. For the "as-isolated" samples (AI), no gas treatment was applied. Here, AI^{red} denotes an (optimal) enzyme isolation procedure in the presence of 2 mM NaDT, whereas AI^{ox} samples are preparations without any reducing agents (NaCl was used instead of NaDT to adjust the ionic strength of the buffers). Samples were filled into Kapton-covered acrylic-glass sample holders for XAS and frozen in liquid nitrogen. EPR measurements on the XAS samples (not shown) revealed no indications

for significant unspecific iron in the as-isolated reduced CrHydA1 samples as the "rhombic iron" signal at a g-value of ~4 was practically absent. The absence of water-dissolved Fe ions in as-isolated reduced CrHydA1 samples was also likely according to the K-edge spectra, which showed no evidence for Fe–O bonds (see Results).

X-ray Absorption Spectroscopy (XAS). Kα fluorescence-detected XAS spectra at the Fe K-edge were collected at 20 K using energy-resolving Ge detectors and helium cryostats as previously described (22, 32) at beamline D2 of the EMBL outstation (at HASYLAB, DESY, Hamburg, Germany) and at beamline KMC-1 of BESSY (Berlin, Germany). Harmonic rejection was achieved by detuning of the Si(111) double-crystal monochromators to 50% of their peak intensities. The energy resolution of the used monochromators was ~3 eV at DESY and optimized by appropriate setting of aperture slits and slightly lower, ~4 eV, at BESSY due to the beamline specifics (33). The slightly different energy resolution of the XAS spectra does not affect the conclusions drawn from the data analysis. Spectra were collected maximally for a scan range of 6950–8450 eV, i.e., up to $k = 19.5$ Å^{-1}, within ~1 h each. Dead time-corrected XAS spectra were averaged after energy calibration of each scan using the peak at 7112 eV in the first derivative of the absorption spectrum of an Fe foil as an energy standard (estimated accuracy ±0.1 eV) (18). EXAFS oscillations were normalized and extracted as described in refs 29 and 34. The energy scale of EXAFS spectra was converted to the wave vector scale (k-scale) using an E_0 value of 7112 eV. Unfiltered k^3-weighted spectra were used for least-squares curve fitting employing a curved-wave multiple-scattering (MS) approach with the program EXCURV (35). In our hands, the EXCURV program is able to reproduce the expected coordination numbers of Fe–ligand interactions in Fe model compounds with known structure from fitting of respective EXAFS oscillations ranging up to k-values of 20 Å^{-1} (18). An amplitude reduction factor, S_0^2, of 0.9 was used in the EXAFS fits. EXAFS fits initially were based on a model of the H-cluster derived from the crystal structures (8, 9). The CO and CN ligands were classified as units, with almost linear arrangement of the Fe–C≡O/N atoms (bond angles close to 180°). For the units, correlation was enabled. A maximal path length in MS calculations of 10 Å and extending over a maximal scattering order of 4 was employed. Debye–Waller factors ($2\sigma^2$) were either fixed at physically reasonable values (see Table 2) or varied in the simulations. Fourier transforms (FTs) were calculated from k^3-weighted EXAFS data using the program SimX (29) and employing $\cos^2$ windows ranging over 10% at both ends of the k-range. From experimental K-edge spectra the preedge peak region was extracted by subtraction of a polynomial spline through the main edge rise using the program XANDA (36). K-edge energies were determined by the "integral method" (29), using integration limits of 15% and 90% of normalized intensity.

RESULTS

As-Isolated Reduced State of CrHydA1. Purification of the protein under strictly anaerobic and reducing conditions (30) to obtain the as-isolated state of CrHydA1 (AI^{red}) yielded the most active protein (Table 1). Iron XAS spectra of AI^{red} were collected up to a k-value of 19.5 Å^{-1}, i.e., to ~1450 eV above the Fe K-edge at ~7120 eV (Figure 1). This very long k-range for XAS on proteins allows for resolution even of closely spaced interatomic distances (34, 37).

RESULTS

Table 1: Specific H_2-Evolving Activities of CrHydA1 Preparations

sample	specific H_2 evolving activity (μmol of H_2 mg^{-1} min^{-1}) $\pm$ 20
CrHydA1^{red}	650
CrHydA1^{ox}	590
H_2 treated	620
CO treated	40

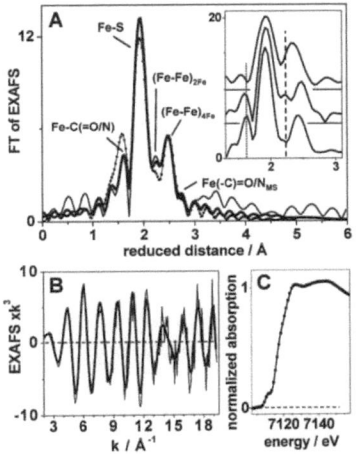

FIGURE 1: XAS analysis of as-isolated reduced CrHydA1 (AI^{red}) at the Fe K-edge. (A) Fourier transform of the EXAFS spectrum in (B). The FT was calculated from k-values ranging over 4.7–19.5 Å^{-1}. Inset: FTs calculated from k-ranges starting at 7.4, 4.9, and 1.6 Å^{-1} (top to bottom); vertical lines mark contributions from Fe–C(=O/N) vectors (dots) and from the Fe–Fe vector in 2Fe$_H$ (dashes). (A) and (B): thin lines, experimental data; thick lines, simulations (Table 2C); the dotted line in (A) is a simulation with parameters in Table 2A. (C) Fe K-edge spectra of two AI^{red} preparations measured at DESY (line) and BESSY (dots).

The K-edge spectrum of AI^{red} (Figure 1C) was perfectly reproducible in independent protein samples, indicating the high quality of the CrHydA1 preparation. Its small primary maximum at ~7130 eV revealed the predominant coordination of Fe by sulfur ligands (22, 38) as expected due to the presence of the respective conserved cysteines in CrHydA1. Indications for oxygen ligation of Fe, as expected for oxidatively modified (or damaged) FeS species (22), were absent.

The Fourier transform (FT) of the AI^{red} EXAFS spectrum shows three major peaks (Figure 1A), at reduced distances between ~1.5 and ~2.5 Å (the reduced distance is the true interatomic distance minus ~0.4 Å due to a phase shift). This immediately suggests at least three Fe–backscatterer shells, i.e., of Fe–C(=O/N), Fe–S, and Fe–Fe vectors. A shoulder on the high-distance side of the FT peak due to Fe–Fe interactions likely accounted for the expected Fe(C)=O/N interactions, the contributions of which to the EXAFS spectrum usually are enhanced by multiple scattering (MS) effects due to the almost linear Fe–C=O/N arrangement (see below). The FT peaks due to Fe–C(=O/N) and Fe(–C)=O/N interactions may be

expected to be similarly large (18). Here, the magnitude of the peak due to Fe(–C)=O/N apparently was diminished due to partial cancelation of respective EXAFS oscillations by interference with EXAFS contributions from the Fe–Fe interactions. A fourth peak at ~2.2 Å reduced distance became more prominent when the FT was calculated starting at higher k-values (Figure 1A, inset) where metal–metal interactions dominate the EXAFS oscillations (shown in Figure 1B). This result was suggestive of a Fe–Fe distance of 2.5–2.6 Å in the H-cluster. The Fe–Fe distance of the 2Fe$_H$ unit in crystal structures of bacterial [FeFe] hydrogenases is about 2.5 Å, whereas the Fe–Fe distances in the [4Fe4S] cluster are ~2.7 Å (8, 9). Thus, the EXAFS data apparently allow to discriminate between the Fe–Fe distances in the [4Fe4S] and the 2Fe$_H$ subgroups of the C. reinhardtii H-cluster.

Precise Fe–ligand distances in AI^{red} were determined by simulations (curve fitting) of the EXAFS spectrum. We used a multiple scattering (MS) approach (35, 39) to account for the expected (relatively small) MS contributions to the EXAFS from the linear CO and CN ligands at 2Fe$_H$. It should be noted that very similar fit results with respect to bond lengths and coordination numbers were obtained if MS contributions very neglected in the simulation procedure (not documented). First, coordination numbers were employed in the simulations (Table 2) which where based on the crystal structures of the H-cluster in bacterial [FeFe] hydrogenases (8, 9, 14). Possible weak Fe–Fe interactions between the two parts of the H-cluster were neglected in the simulations as all distances between Fe ions of the [4Fe4S] cluster and of 2Fe$_H$ were $\geq$4 Å in crystal structures of the bacterial enzymes and such interactions did not contribute significantly to the EXAFS spectra. The fit results in Table 2A and the corresponding fit curve (Figure 1A, dotted line) revealed that the main features of the EXAFS spectrum already were reasonably well reproduced by using the crystal structure as a template. All Fe–ligand distances, in particular the two Fe–Fe and Fe–S distances (Table 2A), are similar to those found in the crystal structures of CpI and DdH (8, 9). Sequence similarity of the catalytic H-domain of these bacterial hydrogenases to CrHydA1 is reasonable high. Still, structures derived from X-ray crystallography, especially for metal cofactors, have to be considered with care. The good agreement of XAS analysis and crystal structure, however, suggests that the overall organization of the green algae H-cluster is similar to that of bacteria.

A more elaborated fit approach yielded refined structural parameters of the H-cluster. Allowing the coordination numbers (N_i) of the Fe–C(=O/N) and Fe–Fe distances to vary in the fit yielded a lower $N_{Fe-C(=O/N)}$ value but a larger N-value of the ~2.5 Å Fe–Fe distance (Table 2B). An $N_{Fe-C(=O/N)}$ value of less than unity, i.e., less than one C(=O/N) ligand per each of the six Fe atoms on the average, was expected for the absence of a CO in a bridging position between the two iron atoms of 2Fe$_H$. In the literature, the reduced form of the H-cluster (H$_{red}$) has been reported to lack this bridging CO, and the respective CO group is bound to Fe$_p$ (p = proximal) solely. Upon oxidation, the CO bridge between the distal and proximal Fe atoms is formed (14, 16, 17, 41–43). The as-isolated AI^{red} protein of CrHydA1 from our preparations is not likely to contain oxidized or CO-inhibited impurities. Due to our fast and gentle isolation protocol, EPR spectra of AI^{red} did not show the axial EPR signal with g-values of 2.102, 2.040, and 1.998 as expected for CrHydA1^{ox} and 2.052, 2.007, and 2.007 for CrHydA1^{ox}-CO as reported recently (27). Roseboom et al. explained the frequent

RESULTS

emergence of CO-inhibited species by cannibalization of free CO from denatured protein (17). However, these impurities were not detected in Al^{red}. A larger apparent coordination number than the expected one of 0.33 (due to two Fe−Fe distances of ~2.5 Å per six Fe ions) of the Fe−Fe distance attributed to 2Fe$_H$ may suggest contributions from similarly long Fe−S vectors. In the structure of reduced [FeFe] hydrogenase from *C. pasteurianum* the distance of Fe$_p$ at the 2Fe$_H$ site to the cysteine sulfur atom which links it to the [4Fe4S] cluster is ~2.50 Å, and the Fe−Fe distance in 2Fe$_H$ is only slightly longer, ~2.55 Å. Thus, such a long Fe−S distance may contribute to the apparent coordination number of the Fe−Fe interaction. We note that in the crystal structures all further Fe−S bonds are shorter than about 2.4 Å, and such distances would not interfere with the 2.5 Å feature.

The coordination number of the Fe−Fe distance of ~2.7 Å ($N_{2.7}$) was very close to 2 (2.02) if it was allowed to vary in the fit (Table 2). This value is similar to the one of 2.0 that is calculated for the presence of only the H-cluster in *Cr*HydA1 (i.e., for 12 Fe−Fe distances of ~2.7 Å in the cubane moiety per 6 Fe ions). In the [FeFe] hydrogenases of *Cp*I and *Dd*H, besides the H-cluster, four and two additional FeS clusters are found so that the $N_{2.7}$ values would be expected to be significantly larger, 2.27 and 2.52, respectively. Thus, our XAS data suggest, in concert with previously performed EPR experiments (27, 30) and multiple sequence alignments (23−25), that only the H-cluster is present in *Cr*HydA1 and further FeS clusters are absent.

A significant further improvement of the fit quality (R_F less than 10%) was achieved by including separate shells for the Fe−C(=O) and Fe−C(=N) interactions and further slight parameter adjustments (Table 2C; Figure 1A,B, thick lines). Now, individual distances for Fe−C(=O) of 1.77 Å and Fe−C (=N) of 1.98 Å were obtained. These distances are close to those of the respective ligands in the crystal structures, where the Fe−C (=O) distance usually is shorter than Fe−C(=N) (8, 9).

We note that the above used "top-down" simulation approach, i.e., starting the EXAFS simulations with a model based on the crystal structures of the H-cluster in bacteria and then refining the structural parameters by additional degrees of freedom in the fit, leads to the same simulation results that were obtained by an inverse ("bottom-up") procedure where the structural model was developed by the stepwise inclusion of increasing numbers of coordination shells in the fit (not documented).

An unusual bridging ligand (which perhaps is an azadithiolate (adt)) has been proposed to be present in the 2Fe$_H$ part of the bacterial H-cluster and may serve crucial functions in H$_2$ production (9, 14, 44−47). The respective carbon and nitrogen or oxygen atoms would be within a range of about 3−4 Å to the Fe atoms of 2Fe$_H$. As such distances overlap with those from Fe (−C)=O/N interactions and only weak contributions from these atoms to the EXAFS spectra were expected, the nature of the bridging species in the *C. reinhardtii* H-cluster cannot be deduced from the XAS data.

H-Cluster Integrity in Protein from Two Purification Conditions. The integrity of the active site was compared in as-isolated *Cr*HydA1 samples purified under reducing (Al^{red}) and mildly oxidizing (Al^{ox}) conditions (Figure 2). The specific H$_2$-evolving activity of Al^{ox} samples, as determined by the in vitro assay, was only slightly lower than in Al^{red} (Table 1). The Fe K-edge was at ~0.5 eV higher energies in Al^{ox} compared to Al^{red} (Figure 2A), indicating that overall more reduced Fe was present in Al^{red}. Both K-edges were at higher energies than that for Fe(II)

Table 2: Structural Parameters of the H-Cluster from EXAFS Simulations^a

sample	shell	fit	N_i (per Fe)	R_i (Å)	$2\sigma_i^2$ (Å^2)	R_F (%)
Al^{red}	C(=O/N)	A	1.00^b	1.78	0.002^b	22.5
	S		2.17^d	2.28	0.001^c	
	S		1.33^d	2.39	0.001^c	
	Fe (2Fe$_H$)		0.33^b	2.52	0.002^b	
	Fe (4Fe4S)		2.00^b	2.73	0.009	
	C(=O/N)	B	0.52	1.77	0.002^b	21.0
	S		2.56	2.29	0.002^b	
	S		1.14	2.40	0.002^b	
	Fe (2Fe$_H$)		0.61	2.53	0.002^b	
	Fe (4Fe4S)		2.02	2.71	0.010	
	C(=O)	C	0.45^e	1.77	0.001^b	8.7
	C(=N)		0.55^e	1.98	0.001^b	
	S		2.04^d	2.28	0.001^b	
	S		1.46^d	2.40	0.003^b	
	Fe (2Fe$_H$)		0.77	2.53	0.002^b	
	Fe (4Fe4S)		2.02	2.72	0.012	
Al^{ox} (Al^{red})	C(=O/N)	D	0.55	1.84	0.001^b	7.7
			(0.74)	(1.86)	(0.001)^b	
	S		2.96	2.28	0.010^b	
			(3.57)	(2.28)	(0.010)^b	
	Fe (2Fe$_H$)		0.58	2.52	0.001^b	
			(0.71)	(2.53)	(0.001)^b	
	Fe (4Fe4S)		1.19	2.70	0.010^b	
			(2.06)	(2.73)	(0.010)^b	
Al^{red} {CO} {H$_2$}	C(=O/N)	E	0.68	1.85	0.001^b	8.3
			[0.92]	[1.79]	[0.001]^b	
			{0.42}	{1.78}	{0.001}^{b}	
	S		3.45	2.29	0.010^b	
			[3.45]^f	[2.27]	[0.010]^b	
			{3.45}^{f}	{2.25}	{0.010}^{b}	
	Fe (2Fe$_H$)		0.54^f	2.53	0.002^b	
			[0.54]^f	[2.62]	[0.002]^b	
			{0.54}^{f}	{2.51}	{0.002}^{b}	
	Fe (4Fe4S)		2.02^f	2.73	0.012^b	
			[2.02]^f	[2.71]	[0.012]^b	
			{2.02}^{f}	{2.71}	{0.012}^{b}	

^{a} N_i, coordination number; R_i, Fe−ligand distance; $2\sigma_i^2$, Debye−Waller factor. ^{b} Fixed values. ^{c} $2\sigma^2$ restricted to values >0. ^{d} The sum of N_S values was 3.5. ^{e} The sum of $N_{C(=O/N)}$ values was 1. ^{f} N_i was coupled to yield equal values for the three spectra. R_F (29) was calculated over reduced distances of 1.3−2.8 Å. All simulations comprise a further multiple-scattering Fe(−C)=O/N shell (with the same N value as for the Fe−C (=O/N) shell; respective distances were in the range of 2.95−3.02 Å; $2\sigma^2$ was set to 0.01 Å^2).

species due to the presence of Fe(III) ions in the cubane cluster (48, 49). The shift of the edge to lower energies in Al^{red} was compatible with the reduction of at least one Fe(III) ion to the Fe(II) state. Furthermore, the preedge peak, due to formally dipole-forbidden 1s → 3d electronic transitions, was increased in Al^{ox} (Figure 2A, inset). This suggests a less symmetric overall Fe coordination (40), presumably due to the binding of oxygen species to some Fe atoms in Al^{ox} (see below). Lowering of the site symmetry is expected to cause increased admixtures of metal p levels to the 3d electronic orbitals, increasing the probability of electronic transitions into unoccupied 3d levels and thus the intensity of the respective preedge feature. Due to the limited energy resolution of the monochromator at BESSY, possible spectral substructure on the preedge peaks, as related, i.e., to electronic multiplet interactions (40, 50), remained invisible.

The FTs of EXAFS spectra revealed a pronounced decrease of the peak due to Fe−Fe interactions in Al^{ox} (Figure 2B, arrow). EXAFS simulations (using a simplified approach) revealed a

RESULTS

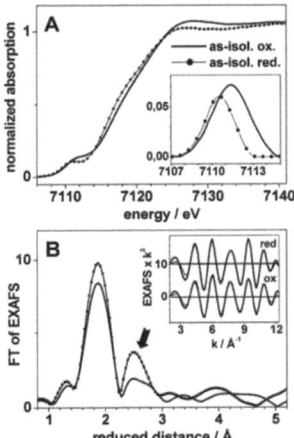

FIGURE 2: XAS comparison of as-isolated reduced (Al^{red}) and oxidized (Al^{ox}) CrHydA1. (A) K-edge spectra (inset, isolated preedge features). (B) FTs of EXAFS spectra (dotted line, Al^{red}; solid line, Al^{ox}). FTs were calculated over a k-range of 1.6−12.5 Å$^{−1}$; the arrow marks the contributions mainly from Fe−Fe interactions. Inset: Fourier isolates over reduced distances of 0.5−3.0 Å (thin lines, experimental data) and simulations according to Table 2D (thick lines).

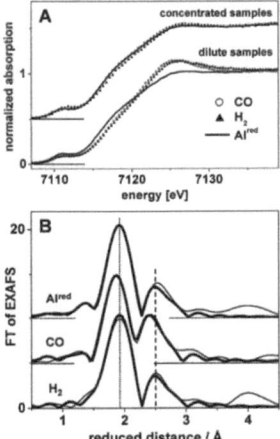

FIGURE 3: Effects of gas treatment on CrHydA1: CO binding and H$_2$ reduction. (A) K-edges of indicated CrHydA1 samples. (B) FTs of EXAFS spectra (thin lines, calculated as in Figure 2) and simulations (thick lines, Table 2E). H$_2$ and CO treatments in (B) were carried out on concentrated protein. Vertical dashed and dotted lines mark shifts in the respective FT peaks due to Fe−S and Fe−Fe interactions.

decrease of the coordination number of ~2.7 Å Fe−Fe vectors by a factor of about 2 and a less pronounced decrease of N_{Fe-S} and $N_{FeC(=O/N)}$. The N-value of the ~2.5 Å Fe−Fe distance in 2Fe$_H$ of Al^{ox}, however, was only slightly smaller than that of Al^{red} (Table 2D). These results may suggest that in Al^{ox} primarily the [4Fe4S] part of the H-cluster was modified or destroyed, whereas in the larger protein fraction the binuclear site at least still contained the ~2.5 Å Fe−Fe distance. Whether this Fe−Fe distance belongs to an otherwise unmodified 2Fe$_H$ cluster cannot be concluded from the data.

Effects of Treatments with CO and H$_2$. The effects of incubation of CrHydA1 with CO and H$_2$ on the structure of the H-cluster strongly depended on whether concentrated or more dilute protein samples were treated (Figure 3). In diluted samples (Figure 3A), the increased primary maxima of the K-edges (at ~7125 eV) and lowered preedge amplitudes (at ~7111 eV) suggested more symmetric coordination of the Fe atoms. Such an effect was expected if Fe ions had become released from the protein to the medium, forming hexaquo−Fe ions. Such deleterious effects in dilute samples were observed with both gases. Analysis of the EXAFS data of H$_2$- and CO-treated dilute protein (not documented) revealed the partial loss of Fe−Fe interactions, which indicates a degradation of the H-cluster.

H$_2$ and CO treatments on more concentrated samples caused only small changes of the K-edge spectra (Figure 3A), suggesting that the integrity of the H-cluster was fully preserved. A slight shift of the edge to higher energies (by ~0.4 eV) in H$_2$-treated CrHydA1 points to the oxidation of one Fe atom at 2Fe$_H$, possibly due to reduction of H-species (8, 18). The respective EXAFS spectrum (Figure 3B) showed only minor changes compared to Al^{red}. The main difference in the structure of the H-cluster, according to EXAFS simulations (Table 2E), was a decreased apparent Fe−C(=O/N) distance and a slightly decreased Fe−Fe distance of 2Fe$_H$ in H$_2$-treated enzyme. Although this small decrease in the Fe−Fe distance may not be significant, it could be caused by the presence of one more oxidized Fe atom (likely Fe$_d$ of 2Fe$_H$; d = distal) due to the binding and reduction of H-species. Notably, the Fe−Fe distance of the [4Fe4S] cluster was not affected by the H$_2$ treatment (Table 2E).

Carbon monoxide is an effective inhibitor of [FeFe] hydrogenases. In a crystallographic structure for H$_{ox}$−CO of CpI, electron density at Fe$_d$ was discussed to be exogenous CO (41). FTIR and EPR analyses support this notion (14, 51). Under certain redox conditions, an endogenous CO is believed to form a "bridge" between the Fe atoms of the 2Fe$_H$ moiety (6, 41). A decreased preedge peak in CO-treated CrHydA1 (Figure 3A) was in agreement with an overall more symmetric Fe coordination upon binding of additional CO to 2Fe$_H$. The EXAFS data (Figure 3B) revealed an increased coordination number of the Fe−C(=O/N) interactions accompanied by enhanced respective multiple scattering contributions to the spectrum and an overall reduced mean FeC(=O/N) distance (Table 2E). These findings indicate the presence of one to two surplus short Fe−C(=O) interactions. In comparison to Al^{red}, the Fe−Fe distance of 2Fe$_H$ was increased significantly by ~0.1 Å. A similar increase of the Fe−Fe distance has been observed in crystal structures of CO-treated bacterial [FeFe] hydrogenase, where one extra CO was found at Fe$_d$ (41). Presumably, in CrHydA1 an extra CO also was bound to Fe$_d$, causing more symmetric (near-octahedral) coordination of this ion.

RESULTS

DISCUSSION

XAS on biological metal centers (BioXAS) is a powerful tool to derive atomic resolution structural information, to determine the electronic configuration, especially in those states which are not accessible by EPR spectroscopy, and to monitor their dynamics during the catalytic reactions (29, 52−55). Here, we used XAS at the Fe K-edge to specifically investigate the catalytic cofactor of biological H_2 turnover in an [FeFe] hydrogenase protein.

The structural parameters from the present XAS study strongly suggest that the overall atomic organization of the H-cluster in the [FeFe] hydrogenase of the green alga *C. reinhardtii* is very similar to that found in crystallographically characterized enzymes from bacteria. Recent EPR investigations performed on CrHydA1 support this notion (27, 30). Furthermore, the XAS data strongly suggest that in CrHydA1 no additional FeS clusters are found, as opposed to bacterial hydrogenases (2, 8, 9). EXAFS analysis allowed precise determination of interatomic distances in the H-cluster. The individual Fe−Fe distances in the [4Fe4S] part and 2Fe$_H$ moiety of the H-cluster were measured (precision on the order of ∼0.02 Å). Changes of the structural parameters upon treatments of CrHydA1 with the inhibitor CO and the substrate H_2 were detected, allowing to address changes, e.g., in catalytic intermediates. The XAS-derived interatomic distances, because of their higher resolution compared to protein crystallography (56), may help to optimize *in silico* structures of the H-cluster in DFT calculations (57−59).

There are also limitations of the XAS analysis. By the applied methods, predominantly average coordination environments of the six Fe atoms in the cluster were obtained. Characterization of the individual structure and oxidation state of the Fe atoms, in particular in the 2Fe$_H$ unit, and of the binding sites of substrate and inhibitors is highly desirable. It may be facilitated by future investigations employing site-selective XAS techniques (60). It has been proposed that an azadithiolate is bridging the Fe atoms of the 2Fe$_H$ site and is essential in H_2 turnover (12, 18, 45−47, 61). The respective C, N, and O atoms are almost impossible to detect by XAS because their distances to Fe are relatively large and overlap with the Fe(C)=O/N distances and with contributions of atoms from the protein backbone. Thus, discrimination between different bridging species likely cannot be obtained by XAS.

The integrity of the algal H-cluster is easily perturbed, as observed for CrHydA1 purified under mildly oxidizing conditions and upon incubation of dilute protein with H_2 and CO. Such conditions seem to cause the release of Fe ions from the protein into the medium and, hence, at least partial degradation of the H-cluster. Preliminary evidence was obtained that the [4Fe4S] cluster is the primary target of oxidative modification, whereas the 2Fe$_H$ moiety may be more robust. O_2 sensitivity of FeS clusters is well-known (62, 63). Purification of CrHydA1 under reducing conditions prevents such deleterious effects and stabilizes an intact H-cluster. In bacterial [FeFe] hydrogenases, the H-cluster is deeply buried in the protein. Induced-fit folding models of CrHydA1 (M. Winkler and T. Happe, unpublished results) suggest that its [4Fe4S] cluster is located just beneath the surface whereas the 2Fe$_H$ moiety faces the inside of the protein. This arrangement is likely to allow for easy access of gas molecules and redox partners from the bulk to the [4Fe4S] unit and for its rapid modification by O_2.

Treatments of CrHydA1 with H_2 and CO revealed relatively subtle but discernible structural changes of the H-cluster. Inter-

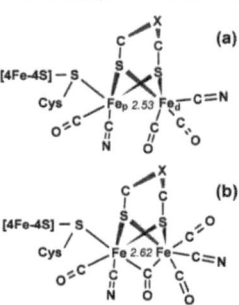

FIGURE 4: Tentative structures of the H-cluster in CrHydA1 for Al^{red} and Al^{ox} CO. (a) As-isolated reduced and (b) CO-treated protein, in agreement with the XAS results (Fe−Fe distances in Å). The bridging ligand on top of the 2Fe$_H$ cluster is (partly) visible in the crystal structures of bacterial [FeFe] hydrogenases (8−12). Its chemical nature, however, is unclear (X may be N or O). The respective C and N or O atoms were not resolved in this XAS study.

action of the substrate H_2 with 2Fe$_H$ seems to cause slight overall oxidation of the two Fe atoms in the 2Fe$_H$ part, but its Fe−Fe distance was almost unchanged. Accordingly, a change in the Fe−Fe bridging motif upon H_2 binding, compared to reduced as-isolated CrHydA1, is unlikely. If H-species become bound to 2Fe$_H$, they may be located in a terminal position on Fe$_d$ but not in an Fe−Fe bridging position. Further investigations are required to clarify this issue. CO binding to 2Fe$_H$, on the other hand, caused a significant elongation of the Fe−Fe distance, and additional short Fe−C(=O) distances were found. These results are comparable with crystal data of CO-treated bacterial [FeFe] hydrogenases (41) and previous EPR data on CrHydA1 (27). Presumably, the extra CO on Fe$_d$ inhibits CrHydA1 by causing a structural rearrangement, which blocks a binding site for H-species, in a similar fashion to that proposed for bacterial enzymes (10, 11, 41).

In conclusion, we show that by XAS on algal [FeFe] hydrogenase the site geometry and electronic configuration of the H-cluster can be resolved. It is possible to monitor changes upon substrate and inhibitor binding. For the first time, the bond lengths and Fe−Fe distances of the H-cluster were determined at subangstrom resolution. We summarize our results on the structure of the H-cluster in as-isolated reduced and CO-treated CrHydA1 in the tentative models shown in Figure 4. Further investigations to characterize the active site of *C. reinhardtii* [FeFe] hydrogenase in its H_2 turnover cycle are in progress in our laboratories.

ACKNOWLEDGMENT

We thank the beamline scientists Dr. W. Meyer-Klaucke (EMBL at DESY, Hamburg) and Dr. F. Schäfers and M. Mertin (BESSY, Berlin) for excellent technical support and Prof. Holger Dau (FU-Berlin) for providing kind access to XAS measuring equipment at BESSY.

REFERENCES

1. Cammack, R., Robson, R., and Frey, M., Eds. (1997) *Hydrogen as a fuel: Learning from nature*, Taylor and Francis, London, U.K.

RESULTS

2. Vignais, P. M., and Billoud, B. (2007) Occurrence, classification, and biological function of hydrogenases: An overview. *Chem. Rev. 107*, 4206–4272.
3. Mertens, A., and Liese, A. (2004) Biotechnological applications of hydrogenases. *Curr. Opin. Biotechnol. 15*, 343–348.
4. Rittmann, B. E. (2008) Opportunities for renewable bioenergy using microorganisms. *Biotechnol. Bioeng. 100*, 203–212.
5. Evans, D. J., and Pickett, C. J. (2003) Chemistry and the hydrogenases. *Chem. Soc. Rev. 32*, 268–275.
6. Adams, M. W. W. (1990) The structure and mechanism of iron-hydrogenases. *Biochim. Biophys. Acta 1020*, 115–145.
7. Frey, M. (2002) Hydrogenases: hydrogen-activating enzymes. *ChemBioChem 3*, 153–160.
8. Peters, J. W., Lanzilotta, W. N., Lemon, B., and Seefeldt, L. C. (1998) X-ray crystal structure of the Fe-only hydrogenase (CpI) from *Clostridium pasteurianum* to 1.8 Å resolution. *Science 282*, 1853–1858.
9. Nicolet, Y., Piras, C., Legrand, P., Hatchikian, C. E., and Fontecilla-Camps, J. C. (1999) *Desulfovibrio desulfuricans* iron hydrogenase: the structure shows unusual coordination to an active site Fe binuclear center. *Struct. Folding Des. 7*, 13–23.
10. Peters, J. W. (1999) Structure and mechanism of iron-only hydrogenases. *Curr. Opin. Struct. Biol. 9*, 670–676.
11. Nicolet, Y., Cavazza, C., and Fontecilla-Camps, J. C. (2002) Fe-only hydrogenases: structure, function and evolution. *J. Inorg. Biochem. 91*, 1–8.
12. Nicolet, Y., Lemon, B. J., Fontecilla-Camps, J. C., and Peters, J. W. (2000) A novel FeS cluster in Fe only hydrogenases. *Trends Biochem. Sci. 25*, 138–143.
13. Volbeda, A., Charon, M. H., Piras, C., Hatchikian, E. C., Frey, M., and Fontecilla-Camps, J. C. (1995) Crystal structure of the nickel-iron hydrogenase from *Desulfovibrio gigas*. *Nature (London) 373*, 580–587.
14. Nicolet, Y., de Lacey, A. L., Vernède, X., Fernandez, V. M., Hatchikian, E. C., and Fontecilla-Camps, J. C. (2001) Crystallographic and FTIR spectroscopic evidence of changes in Fe coordination upon reduction of the active site of the Fe-only hydrogenase from *Desulfovibrio desulfuricans*. *J. Am. Chem. Soc. 123*, 1596–1601.
15. Chen, Z., Lemon, B. J., Huang, S., Swartz, D. J., Peters, J. W., and Bagley, K. A. (2002) Infrared studies of the CO-inhibited form of the Fe-only hydrogenase from *Clostridium pasteurianum* I: examination of its light sensitivity at cryogenic temperatures. *Biochemistry 41*, 2036–2043.
16. Pierik, A. J., Hulstein, M., Hagen, W. R., and Albracht, S. P. (1998) A low-spin iron with CN and CO as intrinsic ligands forms the core of the active site in [Fe]-hydrogenases. *Eur. J. Biochem. 258*, 572–578.
17. Roseboom, W., De Lacey, A. L., Fernandez, V. M., Hatchikian, E. C., and Albracht, S. P. J. (2006) The active site of the [FeFe]-hydrogenase from *Desulfovibrio desulfuricans*. II. Redox properties, light sensitivity and CO-ligand exchange as observed by infrared spectroscopy. *J. Biol. Inorg. Chem. 11*, 102–118.
18. Löscher, S., Schwartz, L., Stein, M., Ott, S., and Haumann, M. (2007) Facilitated hydride binding in an Fe-Fe hydrogenase active-site biomimic revealed by X-ray absorption spectroscopy and DFT calculations. *Inorg. Chem. 46*, 11094–11105.
19. Fontecilla-Camps, J. C., Volbeda, A., Cavazza, C., and Nicolet, Y. (2007) Structure/function relationships of [NiFe]- and [FeFe]-hydrogenases. *Chem. Rev. 107*, 4273–4303.
20. Pereira, A. S., Tavares, P., Moura, I., Moura, J. J., and Huynh, B. H. (2001) Mössbauer characterization of the iron-sulfur clusters in *Desulfovibrio vulgaris* hydrogenase. *J. Am. Chem. Soc. 123*, 2771–2782.
21. Schwab, D. E., Tard, C., Brecht, E., Peters, J. W., Pickett, C. J., and Szilagyi, R. K. (2006) On the electronic structure of the hydrogenase H-cluster. *Chem. Commun. 35*, 3696–3698.
22. Buhrke, T., Löscher, S., Lenz, O., Schlodder, E., Zebger, I., Andersen, L. K., Hildebrandt, P., Meyer-Klaucke, W., Dau, H., Friedrich, B., and Haumann, M. (2005) Reduction of unusual iron-sulfur clusters in the H$_2$-sensing regulatory Ni-Fe hydrogenase from *Ralstonia eutropha* H16. *J. Biol. Chem. 280*, 19488–19495.
23. Happe, T., and Naber, J. D. (1993) Isolation, characterization and N-terminal amino acid sequence of hydrogenase from the green alga *Chlamydomonas reinhardtii*. *Eur. J. Biochem. 214*, 475–481.
24. Happe, T., and Kaminski, A. (2002) Differential regulation of the Fe-hydrogenase during anaerobic adaptation in the green alga *Chlamydomonas reinhardtii*. *Eur. J. Biochem. 269*, 1022–1032.
25. Winkler, M., Heil, B., Heil, B., and Happe, T. (2002) Isolation and molecular characterization of the [Fe]-hydrogenase from the unicellular green alga *Chlorella fusca*. *Biochim. Biophys. Acta 1576*, 330–334.
26. Florin, L., Tsokoglou, A., and Happe, T. (2001) A novel type of Fe-hydrogenase in the green alga *Scenedesmus obliquus* is linked to the photosynthetic electron transport chain. *J. Biol. Chem. 276*, 6125–6132.
27. Kamp, C., Silakov, A., Winkler, M., Reijerse, E. J., Lubitz, W, and Happe, T. (2008) Isolation and first EPR characterization of the [FeFe]-hydrogenases from green algae. *Biochim. Biophys. Acta 1777*, 410–416.
28. Girbal, L., von Abendroth, G., Winkler, M., Benton, P. M., Meynial-Salles, I., Croux, C., Peters, J. W., Happe, T., and Soucaille, P. (2005) Homologous and heterologous overexpression in *Clostridium acetobutylicum* and characterization of purified clostridial and algal Fe-only hydrogenases with high specific activities. *Appl. Environ. Microbiol. 71*, 2777–2781.
29. Dau, H., Liebisch, P., and Haumann, M. (2003) X-ray absorption spectroscopy to analyze nuclear geometry and electronic structure of biological metal centers-potential and questions examined with special focus on the tetra-nuclear manganese complex of oxygenic photosynthesis. *Anal. Bioanal. Chem. 376*, 562–583.
30. von Abendroth, G., Stripp, S., Silakov, A., Croux, C., Soucaillec, P., Girbal, L., and Happe, T. (2008) Optimized over-expression of [FeFe] hydrogenases with high specific activity in *Clostridium acetobutylicum*. *Int. J. Hydrogen Energy 33*, 6076–6081.
31. Wiesenborn, D. P., Rudolph, F. B., and Papoutsakis, E. T. (1988) Thiolase from *Clostridium acetobutylicum* ATCC 824 and its role in the synthesis of acids and solvents. *Appl. Environ. Microbiol. 54*, 2717–2722.
32. Barra, M., Haumann, M., Loja, P., Krivanek, R., Grundmeier, A., and Dau, H. (2006) Intermediates in assembly by photoactivation after heat-induced disassembly of the manganese complex of photosynthetic water oxidation. *Biochemistry 45*, 14523–14532.
33. Schäfers, F., Mertin, M., and Gorgoi, M. (2007) KMC-1: a high resolution and high flux soft x-ray beamline at BESSY. *Rev. Sci. Instrum. 78*, 1–14.
34. Dau, H., Liebisch, P., and Haumann, M. (2004) The structure of the manganese complex of photosystem II in its dark-stable S$_1$-state: EXAFS results in relation to recent crystallographic data. *Phys. Chem. Chem. Phys. 6*, 4781–4792.
35. Tomic, S., Searle, B. G., Wander, A., Harrison, M. N., Dent, A. J., Mosselmans, J. F. W., and Inglesfield, J. E. (2005) New tools for the analysis of EXAFS: The DL_EXCURV package, CCLRC Technical Report DL-TR-2005-001, ISSN 1362-0207.
36. Klementiev, K. V. (2005) XANES dactyloscope for Windows, freeware: www.desy.de/~klmn/xanda.html.
37. Gu, W., Jacquamet, L., Patil, D. S., Wang, H. X., Evans, D. J., Smith, M. C., Millar, M., Koch, S., Eichhorn, D. M., Latimer, M., and Cramer, S. P. (2003) Refinement of the nickel site structure in *Desulfovibrio gigas* hydrogenase using range-extended EXAFS spectroscopy. *J. Inorg. Biochem. 93*, 41–51.
38. Löscher, S., Burgdorf, T., Zebger, I., Hildebrandt, P., Dau, H., Friedrich, B., and Haumann, M. (2006) Bias from H$_2$ cleavage to production and coordination changes at the Ni-Fe active site in the NAD$^+$-reducing hydrogenase from *Ralstonia eutropha*. *Biochemistry 45*, 11658–11665.
39. Zabinsky, S. I., Rehr, J. J., Aukudinov, A., Albers, R. C., and Eller, M. J. (1995) Multiple-scattering calculations of x-ray-absorption spectra. *Phys. Rev. B 52*, 2995–3009.
40. Westre, T. E., Kennepohl, P., DeWitt, J. G., Hedman, B., Hodgson, K. O., and Solomon, E. I. (1997) A multiplet analysis of the Fe K-edge 1s→3d pre-edge features of iron complexes. *J. Am. Chem. Soc. 119*, 6297–6314.
41. Lemon, B., and Peters, J. W. (1999) Binding of exogenously added carbon monoxide at the active site of the iron-only hydrogenase (CpI) from *Clostridium pasteurianum*. *Biochemistry 38*, 12969–12973.
42. De Lacey, A., Stadler, C., Cavazza, C., Hatchikian, E. C., and Fernandez, V. M. (2000) FTIR characterization of the active site of the Fe-hydrogenase from *Desulfovibrio desulfuricans*. *J. Am. Chem. Soc. 122*, 11232–11233.
43. Tye, J. W., Darensbourg, M. Y., and Hall, M. B. (2008) Refining the active site structure of iron-iron hydrogenase using computational infrared spectroscopy. *Inorg. Chem. 47*, 2380–2388.
44. Pandey, A. S., Harris, T. V., Giles, L. J., Peters, J. W., and Szilagyi, R. K. (2008) Dithiomethylether as a ligand in the hydrogenase H-cluster. *J. Am. Chem. Soc. 130*, 4533–4540.
45. Fan, H.-J., and Hall, M. B. (2001) A capable bridging ligand for Fe-only hydrogenase: density functional calculations of a low-energy route for heterolytic cleavage and formation of dihydrogen. *J. Am. Chem. Soc. 123*, 3828–3829.

RESULTS

46. Schwartz, L., Eilers, G., Eriksson, L., Gogoll, A., Lomoth, R., and Ott, S. (2006) Iron hydrogenase active site mimic holding a proton and a hydride. *Chem. Commun.* 2006, 520–522.
47. Henry, R. M., Shoemaker, R. K., DuBois, D. L., and DuBois, M. R. (2006) Pendant bases as proton relays in iron hydride and dihydrogen complexes. *J. Am. Chem. Soc. 128*, 3002–3010.
48. Popescu, V. C., and Münck, E. (1999) Electronic structure of the H cluster in [Fe]-hydrogenases. *J. Am. Chem. Soc. 121*, 7877–7884.
49. Albracht, S. P. J., Roseboom, W, and Hatchikian, E. C. (2006) The active site of the [FeFe]-hydrogenase from *Desulfovibrio desulfuricans*. I. Light sensitivity and magnetic hyperfine interactions as observed by electron paramagnetic resonance. *J. Biol. Inorg. Chem. 11*, 88–101.
50. Voevodskaya, N., Lendzian, F., Sanganas, O., Grundmeier, A., Gräslund, A., and Haumann, M. (2008) Characterization of redox intermediates in the O_2-activating Mn-Fe site in R2 protein of *Chlamydia trachomatis* ribonucleotide reductase by X-ray absorption spectroscopy and EPR. *J. Biol. Chem.* (Epub doi:10.1074/jbc.M807190200).
51. Silakov, A., Reijerse, E. J., Albracht, S. P. J., Hatchikian, E. C., and Lubitz, W. (2007) The electronic structure of the H-cluster in the [FeFe]-hydrogenase from *Desulfovibrio desulfuricans*: A Q-band ^{57}Fe-ENDOR and HYSCORE study. *J. Am. Chem. Soc. 129*, 11447–11458.
52. Strange, R. W., and Feiters, M. C. (2008) Biological X-ray absorption spectroscopy (BioXAS): a valuable tool for the study of trace elements in the life sciences. *Curr. Opin. Struct. Biol. 18*, 609–616.
53. Haumann, M., Liebisch, P., Müller, C., Barra, M., Grabolle, M., and Dau, H. (2005) Photosynthetic O_2 formation tracked by time-resolved x-ray experiments. *Science 310*, 1019–1021.
54. Kleifeld, O., Frenkel, A., Martin, J. M. L., and Sagi, I. (2003) Active site electronic structure and dynamics during metalloenzyme catalysis. *Nat. Struct. Biol. 10*, 98–103.
55. Haumann, M., Grundmeier, A., Zaharieva, I., and Dau, H. (2008) Photosynthetic water oxidation at elevated dioxygen partial pressure monitored by time-resolved X-ray absorption measurements. *Proc. Natl. Acad. Sci. U.S.A. 105*, 17384–17389.
56. Sommerhalter, M., Lieberman, R. L., and Rosenzweig, A. C. (2005) X-ray crystallography and biological metal centers: is seeing believing?. *Inorg. Chem. 44*, 770–778.
57. Fiedler, A. T., and Brunold, T. C. (2005) Computational studies of the H-cluster of Fe-only hydrogenases: geometric, electronic, and magnetic properties and their dependence on the [Fe4S4] cubane. *Inorg. Chem. 44*, 9322–9334.
58. Liu, Z. P., and Hu, P. (2002) A density functional theory study on the active center of Fe-only hydrogenase: characterization and electronic structure of the redox states. *J. Am. Chem. Soc. 124*, 5175–5182.
59. Bruschi, M., Fantucci, P., and De Gioia, L. (2003) Density functional theory investigation of the active site of [Fe]-hydrogenases: effects of redox state and ligand characteristics on structural, electronic, and reactivity properties of complexes related to the [2Fe]H subcluster. *Inorg. Chem. 42*, 4773–4781.
60. Glatzel, P., and Bergmann, U. (2005) High resolution 1s core hole X-ray spectroscopy in 3d transition metal complexes—electronic and structural information. *Coord. Chem. Rev. 249*, 65–95.
61. Lawrence, J. D., Li, H., Rauchfuss, T. B., Benard, M., and Rohmer, M. M. (2001) Diiron azadithiolates as models for the iron-only hydrogenase active site: Synthesis, structure, and stereoelectronics. *Angew. Chem., Int. Ed. 40*, 1768–1771.
62. Beinert, H., Holm, R. H., and Münck, E. (1997) Iron-sulfur clusters: Nature's modular, multipurpose structures. *Science 277*, 653–659.
63. Imlay, J. A. (2006) Iron-sulphur clusters and the problem with oxygen. *Mol. Microbiol. 59*, 1073–1082.

RESULTS

2.4 How Oxygen Attacks [FeFe] Hydrogenases from Photosynthetic Organisms

Sven T. Stripp[*], Gabrielle Goldet[†], Caterina Brandmayr[†], Oliver Sanganas[‡], Kylie A. Vincent[†], Michael Haumann[‡], Fraser Armstrong[†], and Thomas Happe[*§]

[*] Lehrstuhl Biochemie der Pflanzen, AG Photobiotechnologie, Ruhr Universität Bochum, Universitätsstrasse 150, 44801 Bochum, Germany

[†] Inorganic Chemistry Laboratory, University of Oxford, South Park Road, OX1 3QR Oxford, United Kingdom

[‡] Institut für Experimentalphysik, Freie Universität Berlin, Arnimallee 14, 14195 Berlin, Germany

[§] **Corresponding Author:**
Phone: +49-(0)234-32 27026; Fax: +49-(0)234-32 14322,
E-mail: thomas.happe@rub.de

http://dx.doi.org/ 10.1073/pnas.0905343106

RESULTS

How oxygen attacks [FeFe] hydrogenases from photosynthetic organisms

Sven T. Stripp[a], Gabrielle Goldet[b], Caterina Brandmayr[b], Oliver Sanganas[c], Kylie A. Vincent[b], Michael Haumann[c], Fraser A. Armstrong[b], and Thomas Happe[a,1]

[a]Lehrstuhl Biochemie der Pflanzen, AG Photobiotechnologie, Ruhr Universität Bochum, Universitätsstrasse 150, 44801 Bochum, Germany; [b]Inorganic Chemistry Laboratory, University of Oxford, South Parks Road, Oxford OX1 3QR, United Kingdom; and [c]Institut für Experimentalphysik, Freie Universität Berlin, Arnimallee 14, 14195 Berlin, Germany

Edited by Bob B. Buchanan, University of California, Berkeley, CA, and approved August 28, 2009 (received for review May 14, 2009)

Green algae such as *Chlamydomonas reinhardtii* synthesize an [FeFe] hydrogenase that is highly active in hydrogen evolution. However, the extreme sensitivity of [FeFe] hydrogenases to oxygen presents a major challenge for exploiting these organisms to achieve sustainable photosynthetic hydrogen production. In this study, the mechanism of oxygen inactivation of the [FeFe] hydrogenase CrHydA1 from *C. reinhardtii* has been investigated. X-ray absorption spectroscopy shows that reaction with oxygen results in destruction of the [4Fe-4S] domain of the active site H-cluster while leaving the di-iron domain (2Fe$_H$) essentially intact. By protein film electrochemistry we were able to determine the order of events leading up to this destruction. Carbon monoxide, a competitive inhibitor of CrHydA1 which binds to an Fe atom of the 2Fe$_H$ domain and is otherwise not known to attack FeS clusters in proteins, reacts nearly two orders of magnitude faster than oxygen and protects the enzyme against oxygen damage. These results therefore show that destruction of the [4Fe-4S] cluster is initiated by binding and reduction of oxygen at the di-iron domain—a key step that is blocked by carbon monoxide. The relatively slow attack by oxygen compared to carbon monoxide suggests that a very high level of discrimination can be achieved by subtle factors such as electronic effects (specific orbital overlap requirements) and steric constraints at the active site.

EXAFS | H-cluster | protein film electrochemistry | biological hydrogen production | green algae

Hydrogenases are ubiquitous in bacteria and archaea but are also found in some eukaryotes, particularly green algae (1). There are three distinct classes, known as [NiFe]-, [FeFe]-, and [Fe] hydrogenases, based on the metal components of the active site that binds or releases H_2 (2).

Many hydrogenases have extremely high activities (3), a fact that has been emphasized most recently in studies by protein film electrochemistry (4–6). Hydrogenases are able to catalyze both H_2 oxidation and H_2 evolution with minimal electrochemical overpotential (driving force) (4, 7), comparable to the $2H^+/H_2$ equilibrium established on platinum (5). The [FeFe] hydrogenases are of particular interest as they tend to be more biased toward H_2 evolution than [NiFe] hydrogenases (8). The active site of [FeFe] hydrogenases, a complex structure known as the "H-cluster," consists of a binuclear Fe center (2Fe$_H$) linked to a [4Fe-4S] cluster (9). Numerous publications report the chemical synthesis of analogues for the 2Fe$_H$ domain and even the entire H-cluster (10)—such is the interest displayed not only in understanding the enzymes, but also in finding cheap alternatives to Pt catalysts. However, [FeFe] hydrogenases are extremely prone to irreversible inactivation by O_2, and this sensitivity is a key challenge for both the biotechnological and the synthetic chemistry approaches (8).

Viewed in detail, the 2Fe$_H$ domain consists of iron atoms Fe$_p$ and Fe$_d$ that are, respectively, proximal and distal relative to the [4Fe-4S] domain that is connected to Fe$_p$ by a bridging cysteine sulfur (9). An unusual dithiolate ligand, originally modeled as a 1,3 propane dithiolate, forms a bridge between Fe$_p$ and Fe$_d$. In the oxidized state H_{ox}, as determined from the structure of the *Cp*I enzyme from *Clostridium pasteurianum*, Fe$_p$ is also coordinated by one CO and one CN^- ligand and shares a bridging CO with Fe$_d$ (11, 12). The distal Fe is also coordinated by one CO and one CN^- ligand, and an additional binding site is vacant (9) or occupied by an exchangeable O ligand, most likely a water molecule (11). In the structure of the [FeFe] hydrogenase from *Desulfovibrio desulfuricans*, which is believed to be crystallized in the H_{red} form, the bridging CO is replaced by a terminal CO on Fe$_d$ (13). A recent EPR analysis favors an oxidation state assignment of $[4Fe-4S]^{2+}$-Fe$_p$(I)Fe$_d$(II) for H_{ox}, with some spin density delocalized onto the [4Fe-4S] domain (14). The EPR-silent H_{red} state is assigned as $[4Fe-4S]^{2+}$-Fe$_p$(I)Fe$_d$(I) or a hydrido species $[4Fe-4S]^{2+}$-Fe$_p$(II)Fe$_d$(II)-H$^-$, and it appears that the [4Fe-4S] cluster may not access the 1+ level in anything other than a transient manner (15). Crystallography and infrared spectroscopy has shown that exogenous CO, a competitive inhibitor, attacks 2Fe$_H$ at the vacant/exchangeable binding site of Fe$_d$ (16–18). Carbon monoxide is a π-acceptor ligand and binds to electron-rich transition metals (19). From a molecular orbital perspective, H_2 resembles CO because, by analogy with the Dewar-Chatt-Duncanson model for binding of alkenes to metals (20), back donation of electron density into the antibonding σ-orbital of molecular H_2 is important for its binding and activation (20). These facts are highly relevant because they form the basis for CO being competitive with H_2 (21) during H_2 oxidation, and therefore, Fe$_d$ is likely to be the site for H_2 binding (17). Like CO, O_2 is also a π-acceptor ligand and likely to bind to the same site(s); we can anticipate that any such binding could result in the generation of highly reactive oxygen intermediates (22, 23).

Here, we present a study of the mechanism by which O_2 irreversibly attacks the H-cluster, by using electrochemical kinetics with the *reversible* inhibitor CO as a complementary probe. By using X-ray absorption spectroscopy (XAS) at the Fe K-edge, we examine, at atomic level resolution, the nature of the product obtained. Our subject is the [FeFe] hydrogenase CrHydA1 from the photosynthetic green alga *Chlamydomonas reinhardtii*. The [FeFe] hydrogenases of green algae are of particular interest for photosynthetic H_2 production (8) and because this class of hydrogenase enzymes contain *only* the H-cluster (24–26) it is possible to interpret the XAS data without interference from additional electron-transferring FeS clusters that are present in bacterial hydrogenases (9, 11). The results

---

Author contributions: S.T.S., G.G., K.A.V., M.H., F.A.A., and T.H. designed research; S.T.S., G.G., C.B., O.S., and M.H. performed research; S.T.S., G.G., C.B., O.S., K.A.V., M.H., F.A.A., and T.H. analyzed data; and S.T.S., G.G., M.H., F.A.A., and T.H. wrote the paper.

The authors declare no conflict of interest.

This article is a PNAS Direct Submission.

[1]To whom correspondence should be addressed. E-mail: thomas.happe@rub.de.

This article contains supporting information online at www.pnas.org/cgi/content/full/0905343106/DCSupplemental.

RESULTS

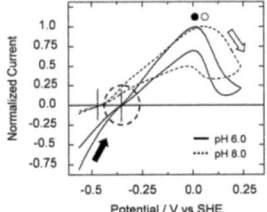

Fig. 1. Catalytic profile of CrHydA1 at pH 6.0 (solid lines) and pH 8.0 (dashed lines) as viewed by cyclic voltammograms of the enzyme adsorbed on a PGE electrode. The thermodynamic $2H^+/H_2$ potentials at pH 6.0 and pH 8.0 are marked by the vertical lines. The dashed oval highlights the inflection point at the zero-current potential at pH 6.0. Black and open circles mark the potential at which anaerobic inactivation begins to occur as the potential is swept to more positive values and the current at each cyclic voltammogram has been normalized at this potential. Black and open arrows indicate the directions of the scans at pH 6 and pH 8, respectively. Experimental conditions: 20 °C, 1 bar H_2, electrode rotation rate 3,000 rpm, scan rate 20 mV/s.

lead to a mechanistic model of how [FeFe] hydrogenases are inactivated by O_2.

Results

Protein Film Electrochemistry. Cyclic voltammograms recorded for *C. reinhardtii* [FeFe] hydrogenase CrHydA1 at 1 bar H_2 show the relative activities for H^+ reduction and H_2 oxidation (Fig. 1). The graph displays two distinct scans obtained at pH 6.0 and pH 8.0 at 20 °C. The potential is swept from -0.55 V to $+0.24$ V vs. SHE and then scanned back to -0.55 V. The enzyme is clearly a bidirectional hydrogenase; at pH 6.0 (solid line), the catalytic activities for H^+ reduction compared to H_2 oxidation are approximately comparable, whereas at pH 8.0 (dashed line), the activity for H^+ reduction is much lower, most obviously because the H^+ concentration is two orders of magnitude lower. In either case the voltammograms cut through the zero-current line (potential axis) at the potential expected for the $2H^+/H_2$ redox couple under these conditions. The slight inflection detectable in this region (dashed oval in Fig. 1) suggests that a small overpotential is required for efficient electron transport to and from the active site. At high potentials (> 0 V vs. SHE, see circles) the enzyme inactivates, giving rise to the anaerobically oxidized state [H_{ox}inact (7)]. This is apparent from the decrease in H_2 oxidation current, which recovers at least partially on the return scan. The reversibility of this inactivation depends on pH, and it is much more reversible at pH 6.

The kinetics of O_2 inactivation of H_2 oxidation were investigated by chronoamperometric experiments in which the current was monitored following changes in gas composition. The catalytic current is a direct measure of enzymatic turnover rate. Fig. 2 shows the dependence of inactivation rate on O_2 level (Fig. 2A) and H_2 level (Fig. 2B). In all experiments the cell potential was set to -0.05 V vs. SHE to optimize the H_2 oxidation rate while avoiding anaerobic inactivation (see circles in Fig. 1). Experiments were performed at pH 6.0, 20 °C. Reactions with O_2 were initiated by changing the gas composition flowing in the headspace of the cell and simultaneously injecting a solution of buffer preequilibrated with the desired gas composition. This combination of operations ensured rapid reaction initiation and a constant O_2 level throughout the reaction. An important feature of these experiments was that the H_2 concentration in solution always remained constant. Control experiments were performed to assess the current contribution because of direct reduction of O_2 at the graphite electrode, although this was small at -0.05 V (see *SI Text* and Fig. S1). In all cases O_2 caused almost complete inactivation (>95%).

Fig. 2A shows how the rate of inactivation depends on the concentration of O_2 (0.5%, 5%, and 10%) with 80% H_2 as the carrier gas and N_2 making up the remaining fraction. Each experiment commenced with 80% H_2 and 20% N_2 flushing through the cell, then O_2 was introduced, as explained above, and once the enzyme was fully inactivated, the original headgas composition was restored by changing the incoming gas mixture back to 80% H_2 and 20% N_2. The time-courses conformed to first-order exponential behavior for one to two half lives and rates were also first order in O_2 up to at least 15% O_2 headgas

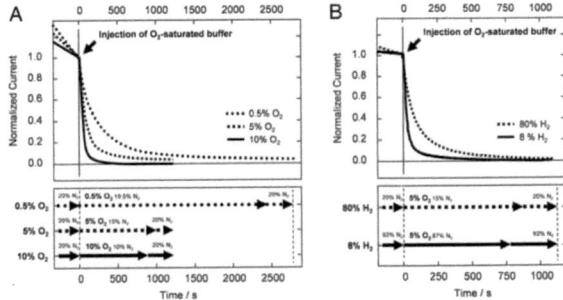

Fig. 2. Inactivation of CrHydA1 by O_2 by simultaneous gas exchange and injection of O_2-saturated buffers. Experiments were carried out under (A) different concentrations of O_2 and (B) different concentrations of H_2. (A) Gas mixtures in the headspace contain 80% H_2 and the remaining 20% are as indicated. For the experiments in which inactivation was induced by 10% and 5% O_2, injections of 2 mL and 0.67 mL buffer saturated with 20% O_2 and 80% H_2 (respectively) were performed into the cell containing 2 mL at the beginning of the experiment. For the experiments in which inactivation was induced by 0.5% O_2, an injection of 0.5 mL of buffer saturated with 2.5% O_2 and 80% H_2 was performed into the cell containing 2 mL at the beginning of the experiment. (B) Inactivation was achieved with 5% O_2 in the headspace and the solid and dashed lines represent experiments performed with 8% and 80% H_2, respectively. For the experiments in which inactivation was induced by 5% O_2 in 8% H_2, an injection of 0.67 mL of buffer saturated with 20% O_2 and 8% H_2 was performed into the cell initially containing 2 mL. Other conditions: pH 6.0, 20 °C, electrode rotation rate 3,000 rpm, -0.05 V vs. SHE.

RESULTS

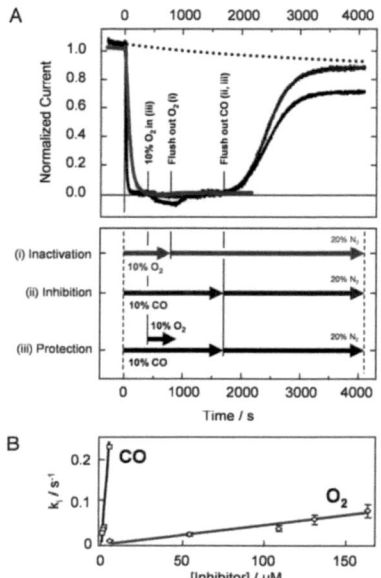

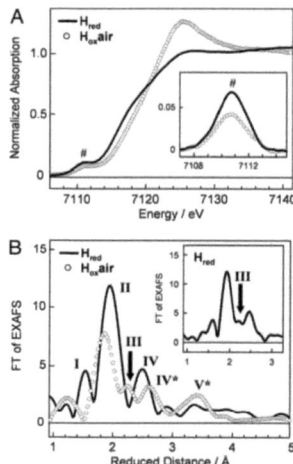

Fig. 3. Inactivation of CrHydA1 by O_2 as compared to inhibition by CO, and protection by CO against O_2 inactivation. (A) Inactivation and inhibition by 10% O_2 (i, red trace) and 10% CO (ii, blue trace) by gas exchange. The black trace shows an experiment in which the enzyme was subjected to 10% O_2 after being fully inhibited by 10% CO (iii). The dotted line gives a visual guide to the progression of background film loss. Experimental conditions: pH 6.0, 20 °C, electrode rotation rate 3,000 rpm, −0.05 V vs. SHE. Gas mixtures in the headspace as indicated, with 80% H_2 and balance of N_2 making up the remainder of the headspace atmosphere. The timeline shown in the lower panel provides a guide for the sequence of gas changes. (B) Dependence of rate of inactivation on concentration of O_2 (diamonds, red) and CO (squares, blue). Note the rates of inactivation by CO were calculated by performing experiments such as those shown in Fig. 2, that is, by simultaneous injection of CO-saturated buffer and gas exchange.

Fig. 4. XAS comparison of CrHydA1 in its as-isolated reduced form (H_{red}) and after O_2 incubation (H_{ox}air). The reduced distance is the true metal-backscatterer distance minus approximately 0.4 Å because of a phase shift. (A) Fe K-edge spectra. Inset: isolated preedge features due to $1s \rightarrow 3d$ electronic transitions. The shown preedge features were derived by subtraction of a polynomial spline from the main edge rise by using the program Xanda. (B) FTs of EXAFS spectra (see Fig. S2) of H_{red} (solid line) and H_{ox}air (open circles). The FTs were calculated for k values of 1.6–16.5 Å^{-1} (7.4–19.5 Å^{-1} in the inset). Numbers on the FT peaks denote specific Fe-ligand interactions as discussed in the text. The Fe-Fe distance of $2Fe_H$ of 2.52 Å (FT peak III) is discernable under both conditions; the Fe-Fe distances of approximately 2.7 Å (FT peak IV) from the [4Fe-4S] cluster are largely diminished in H_{red}.

composition ($\approx$0.2 mM in solution) as displayed in Fig. 3B (see below). Fig. 2B shows that the rate of inactivation of H_2 oxidation by O_2 depends on the H_2 concentration (80% H_2 compared to 8% H_2).

Fig. 3 shows experiments in which CO, known to be a competitive inhibitor of H_2 oxidation (21), is used to probe the course of O_2 attack. Fig. 3A shows three experiments performed at pH 6.0, 20 °C, and −0.05 V vs. SHE. The timeline shown in the lower panel of Fig. 3A provides a guide for the sequence of gas changes. Hydrogen was maintained at 80% fractional composition in the gas stream throughout all experiments. In the first experiment (i), the enzyme is inactivated by introducing 10% O_2 (red trace), and in the second (ii), it is inhibited by introducing 10% CO (blue trace). In the third experiment (iii), 10% O_2 is introduced to enzyme that is already inhibited by 10% CO (black trace).

The decrease in current because of inhibition by CO (ii) is much more rapid than it is for O_2 (i), and the comparative rates of inhibition at different inhibitor concentrations are shown in Fig. 3B. For comparison the second-order rate constants for inhibition by CO and inactivation by O_2 (obtained from the slopes of the lines in Fig. 3B) are 50 mM^{-1}s^{-1} (46 bar^{-1}s^{-1}) and 0.43 mM^{-1}s^{-1} (0.47 bar^{-1}s^{-1}), respectively. Also, it is clear from Fig. 3A that, when the CO is removed from the gas stream, the current increases to a level consistent with that expected based on natural protein film loss from the electrode measured in control experiments over the same period, showing that CO inhibition of H_2 oxidation is reversible. However, removal of O_2 from the cell only results in a tiny increase in current, that is, O_2 inactivation is almost (but not entirely) irreversible. Experiment (iii) in Fig. 3A demonstrates that CO protects the enzyme against inactivation by O_2 as 80% of the initial activity is recovered on removal of CO from the gas stream. This result agrees with the observation reported recently for another [FeFe] hydrogenase CaHydA from Clostridium acetobutylicum (27).

X-Ray Absorption Measurements. Fig. 4 depicts XAS spectra of the as-isolated reduced form of CrHydA1 (H_{red}) and the as-isolated protein after incubation with O_2 (H_{ox}air). The shape of the Fe K-edge spectrum of H_{red} (Fig. 4A) reveals a predominant coordination of the Fe atoms by the S atoms of the conserved cysteine residues in CrHydA1 (25); indications for O-ligation to Fe are absent (28, 29). The increased primary K-edge maximum

RESULTS

Table 1. EXAFS fit results

Sample	Shell	Peak	N_i [per Fe atom]	R_i [Å]	$2\sigma^2_i$ [Å^2]	R_f [%]
H_{red}	C(=O/N)	I	0.46	1.76	0.002*	8.7
	S	II	3.57	2.28	0.010*	
	Fe 2Fe$_H$	III	0.81	2.52	0.002*	
	Fe [4Fe-4S]	IV	2.06	2.71	0.010*	
H_{ox}air	C(=O/N) / O	I	1.11	1.89	0.002*	11.0
	S	II	2.09	2.26	0.010*	
	Fe 2Fe$_H$	III	0.67	2.56	0.002*	
	Fe [4Fe-4S]	IV	0.97	2.77	0.010*	

R_f (39) was calculated over reduced distances of 1.3–2.8 Å. Both simulations comprise a further multiple-scattering Fe(-C)=O/N shell (with the same N-value as for the Fe(=O/N) shell; average respective distances of 2.98 Å; $2s^j$ was set to 0.01 Å^2). N_i, coordination number; R_i, Fe-ligand distance; $2s^2_i$, Debye-Waller factor.
*Fixed values in the fit procedure.

at approximately 7,125 eV and a decreased preedge amplitude (# at ~7,111 eV, inset) in H_{ox}air clearly suggest the binding of O-atom ligands to Fe ions that subsequently became more symmetrically coordinated (25). This result may be explained by binding of additional O-species to the H-cluster and partial release of Fe from the protein in the form of hexaquo-Fe(II) ions.

Fourier transforms (FTs) of EXAFS spectra for H_{red} and H_{ox}air are shown in Fig. 4B. The FT of the H_{red} spectrum (solid line and inset) shows four main peaks I - IV. The reduced distances given in the figure are approximately 0.4 Å smaller than the true metal-ligand distances because of a phase shift. These peaks reflect different Fe interactions of the H-cluster, namely Fe-C(= O/N) (I), Fe-S (II), and Fe-Fe (IV) (25). An additional peak III became visible in H_{red} when the FT was calculated from the corresponding EXAFS oscillations starting at higher k-values and extending over a longer k-range (inset, see Fig. S2). This feature represents the Fe-Fe interaction in the 2Fe$_H$ moiety (25). Peak III in the spectrum of H_{ox}air is also attributable to this Fe-Fe interaction. Diminished Fe-Fe interactions from the cubane cluster in the H_{ox}air spectrum allow for discrimination of peak III even in FT spectra calculated from a k-range starting at lower values. Contributions to the EXAFS spectra from multiple-scattering (MS) effects of the near-linear Fe-C = O/N arrangements are small (25) because of their relatively low coordination number and interference with the Fe-Fe interactions of the [4Fe-4S] cluster.

Precise Fe-ligand distances were determined by simulations of the EXAFS spectra. The coordination numbers per Fe ion (N_i) and the Fe-ligand distances and Fe-Fe distances (R_i) for H_{red} (Table 1) are compatible with the expected structure of the reduced H-cluster H_{red} in CrHydA1. Whereas H_{ox} is reported to be in the Fe$_p$(I)-Fe$_d$(II) state and has a 'bridging' CO between Fe$_p$ and Fe$_d$, H_{red} is assigned as Fe$_p$(I)-Fe$_d$(I) and lacks the bridging CO that has become a terminal ligand to Fe$_d$ (14, 17, 24, 25). In particular, the Fe-Fe distances of 2.52 Å for the 2Fe$_H$ moiety (III, inset in Fig. 4B) and 2.71 Å for the [4Fe-4S] moiety (IV) can clearly be distinguished by EXAFS.

The FT of the EXAFS spectrum of the O_2-treated enzyme (H_{ox}air in Fig. 4B, open circles) reveals a missing contribution from Fe-Fe interactions of the [4Fe-4S] cluster (IV in H_{red}). In addition, a shift of the main FT peak (II) to shorter distances is observed. This is because of contributions from Fe-O bonds, which are shorter than Fe-S bonds. Peak I (representing Fe-C(= O/N) interactions in H_{red}) appears decreased in size in H_{ox}air. This is attributable to interference between Fe-O and Fe-C(= O/N) contributions that cannot easily be discriminated. For H_{ox}air the 2Fe$_H$-specific FT feature III is even more distinct than in H_{red} (inset). Peaks exclusively appearing in the H_{ox}air spectrum are marked with asterisks. Peak IV*, which is indicative of MS contributions from the C = (O/N) ligands is resolved more clearly because peak IV is essentially diminished in H_{ox}air. That the Fe atoms of the [4Fe-4S] cluster may remain bound to the protein but coordinate additional O-ligands is suggested by the observed FT peak V*, which may correspond to long Fe-O-Fe binding motifs (28) prominent in H_{ox}air only.

Simulations of the H_{ox}air EXAFS spectrum (Table 1) revealed that in H_{ox}air the Fe-Fe distance in 2Fe$_H$ is 0.04 Å longer than the Fe-Fe distance in H_{red} (2.56 and 2.52 Å, respectively), with a coordination number lowered by only ~20%. This is compatible with a Fe-Fe distance elongation upon oxidation from H_{red} to H_{ox}-CO which we reported recently (25). Accordingly, a fraction of approximately 80% of protein may be calculated to retain a normal 2Fe$_H$ unit in which the overall structure is preserved upon O_2 treatment. Distance elongation may be caused by binding of O-species to 2Fe$_H$. In contrast, there was a 2-fold decrease of the approximately 2.7 Å Fe-Fe interactions (from N_i 2.06 to 0.97 in H_{ox}air) per Fe ion, because of the loss of respective motifs in the [4Fe-4S] cluster. Therefore, in more than half of the protein molecules the native structure of the [4Fe-4S] cluster is degraded.

In summary, the XAS results show that exposure of CrHydA1 to O_2 causes modification or destruction of the [4Fe-4S] domain of the H-cluster, with the coordination shell of the 2Fe$_H$ domain remaining relatively unchanged.

Discussion

The electrocatalytic response of CrHydA1 shows that the enzyme exhibits very similar characteristics to other hydrogenases (30). In electrochemical experiments, the enzyme's activity is directly measured through the catalytic current. The inflection point at the zero-current potential observed for CrHydA1 contrasts with the sharp intersection with the zero-current axis exhibited by other (bacterial) hydrogenases (6, 7, 27). This may be because of the absence of an electron-transfer relay in CrHydA1 (26) and reflects a small overpotential requirement to drive electrons in either direction. Catalytically, the enzyme is bidirectional and its proficiency in H_2 oxidation allows us to measure *directly* the competition between H_2, CO, and O_2. Note that continuous recording of the time course of H_2 evolution, which requires a low potential or strong reductant, are complicated in the presence of O_2.

Several observations point to a mechanism in which damage by O_2 requires its prior attack at the 2Fe$_H$ domain. First, exogenous CO, known to bind at the distal Fe site (17) and be a competitive inhibitor of H_2 oxidation (21), protects the enzyme against inactivation by O_2. Second, the rate of inactivation by O_2 is lower at high H_2 levels, in line with the competitive relationship between CO and H_2. Third, there are many examples of O_2 and its partial reduction products known as 'reactive oxygen species' (ROS) attacking FeS clusters, including fumarate-nitrate regulatory protein (FNR) (22). Fourth, there is *no* precedent for CO directly attacking [4Fe-4S] clusters in proteins.

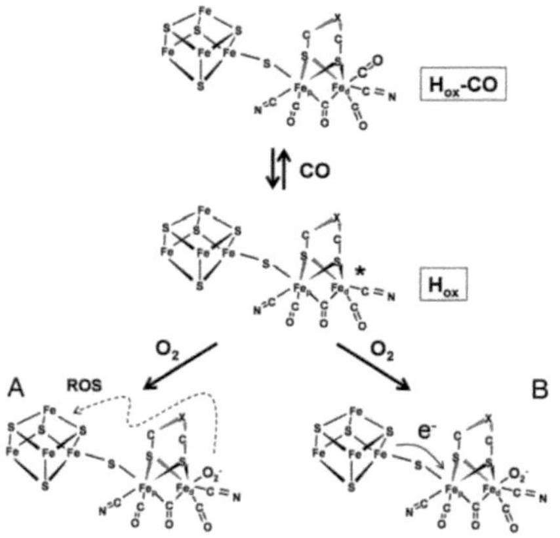

RESULTS

Fig. 5. Model scheme for reaction of CO and O_2 with the H-cluster. Carbon monoxide binds reversibly to the oxidized state H_{ox}, giving an inhibited species H_{ox}-CO. Oxygen reacts by binding to the H-cluster at the same site as CO, that is, Fe_d. The O_2 is converted to a reactive oxygen species (ROS), most likely superoxide (formed by one electron reduction). The ROS can either migrate the very short distance to oxidize the [4Fe-4S]$^{2+}$ cluster (A) or it can remain bound and exert its destructive effect by causing a through-bond electron transfer from the [4Fe-4S] cluster (B).

The observation from EXAFS, that the fate of the H-cluster in its reaction with O_2 is a modification of the [4Fe-4S] domain rather than of the $2Fe_H$ domain, in conjunction with the protective effect of CO, together mean that the reaction of O_2 with the $2Fe_H$ domain (as argued above) generates a species that subsequently attacks the [4Fe-4S] domain. In Fig. 5, we consider two mechanisms by which this could occur. The first option involves O_2 reacting at the $2Fe_H$ domain to form a ROS, particularly superoxide, that migrates the short distance to attack the [4Fe-4S] cluster. The second option involves superoxide being formed by reaction with the $2Fe_H$ domain but remaining bound (presumably at the distal Fe site) as a strong oxidant, inflicting long-range oxidation of the [4Fe-4S]-cluster by through-bond electron transfer. Direct attack of O_2 on the [4Fe-4S] cluster might be prevented by steric effects, or be less favorable, thermodynamically, because superoxide (0.9 V) is a much stronger one-electron oxidant than O_2 (< −0.1 V). In either of these two mechanisms, it is likely that O_2 attacks the H-cluster in the oxidation level H_{ox} that is assigned as [4Fe-4S]$^{2+}$-Fe$_p$(I)Fe$_d$(II). This proposal is based on analogy with observations made with CO, which is known to bind preferentially to H_{ox} relative to H_{red} (giving a characteristic EPR-detectable form known as H_{ox}-CO) (14, 17) [Under anaerobic conditions, it is widely reported that H_{ox} can be oxidized reversibly to the state H_{ox}inact, correspondingly formulated as [4Fe-4S]$^{2+}$-Fe$_p$(II)Fe$_d$(II) (7, 31)]. It is well-established that [4Fe-4S] clusters are prone to oxidative damage via sequences involving some or all of the following: (i) initial formation of superoxidized species such as [4Fe-4S]$^{3+}$ (ii) subsequent ejection of Fe giving a [3Fe-4S] cluster (32–35), (iii) further loss of Fe and S, and (iv) complete breakdown of the cluster. As an example, oxidative breakdown of a [4Fe-4S] cluster by such a sequence is found for FNR, an O_2 sensing protein (22, 36). Oxidative degradation of [4Fe-4S] clusters in ferredoxins without O_2, that is, by an anaerobic pathway, has been studied by cyclic voltammetry and electrochemical potential pulse experiments (34, 35). These studies showed a correlation between the air-stabilities of [4Fe-4S] clusters in different proteins and the electrode potential required to induce their anaerobic oxidative damage via a [3Fe-4S] intermediate. Our observation that anaerobic inactivation of CrHydA1 is only partially reversible may be relevant to the O_2 mechanism because it suggests the ease by which the [4Fe-4S] domain could be damaged by a long-range electron-transfer process, as would be the case if the ROS did not leave the $2Fe_H$ domain.

The observation that CO reacts so much faster than O_2 leads us to question the effectiveness of a 'gas filter' within the enzyme and to suggest instead that more effective gas discrimination occurs in the highly specific region of the active site. In the close proximity of the H-cluster, electronic and steric influences of the protein environment are likely to determine the efficiency of ligand binding. In the case of CrHydA1, we were able to show that binding of CO is kinetically favored about O_2 and H_2 ligation. There is already a precedent for such discrimination, in the reverse direction, in the case of the [NiFe] uptake hydrogenases from *Ralstonia eutropha* and *R. metallidurans*. For those enzymes, which are highly O_2-tolerant, CO (a strong π-acceptor ligand) is a very weak inhibitor (37): even more significantly, in H_2 evolution, H_2 is a far superior inhibitor to either CO or O_2. These effects must arise from differences in the geometries of coordination imposed by the local environment, and we propose that similar principles could be applied to selectivity at the $2Fe_H$ domain.

Materials and Methods

Recombinant *C. reinhardtii* [FeFe] hydrogenase CrHydA1 was synthesized and isolated anaerobically from *C. acetobutylicum* as previously described (38). To prepare samples for XAS, isolated protein was concentrated to 1 mM (48 g/L) by using Vivaspin 6 and Vivaspin 500 columns (Sartorius Stedim Biotech) and stored in 0.1 mM Tris/HCl pH 8.0, 2 mM sodium dithionite (NaDT), and 10% glycerol. To obtain an O_2-treated sample, the protein sample was dialyzed with NaDT-free 0.1 M Tris/HCl pH 8.0 before concentration. Protein samples (30 μL, 1 mM) were placed in 200 μL PCR tubes (Eppendorf) and loaded into 8 mL glass tubes. The tubes were then sealed with a septum and taken out of the anaerobic glovebox (Coy Laboratories, N_2 atmosphere with approximately 5% H_2 and O_2 < 2 ppm). The headgas of the sealed tubes was flushed with moistened O_2 (Air Products, 15 min) and argon (Air Liquide, 5 min) giving the irreversibly damaged form "H_{ox}air". During O_2 treatment, the protein was kept on ice. A sample of CrHydA1 as-isolated and reduced by 2 mM NaDT (26) is referred to as "H_{red}". Samples of H_{ox}air and H_{red} were transferred to Kapton-covered acrylic-glass XAS sample holders and then rapidly frozen in liquid nitrogen until use at the synchrotron.

Protein Film Electrochemistry. Experiments were carried out in an anaerobic glovebox (MBraun) comprising a N_2 atmosphere (O_2 < 2 ppm). Buffers consisting of 0.05 M phosphate with 0.1 M NaCl as additional supporting electrolyte were prepared by using standard reagents NaCl, NaH$_2$PO$_4$, and Na$_2$HPO$_4$ (analytical reagent grade, Sigma). A pyrolytic graphite edge (PGE) rotating disk electrode (RDE, area 0.03 cm^2) was used in conjunction with an electrode rotator (EcoChemie). The all-glass electrochemical cell featured a Pt counter electrode placed in the main compartment and a saturated calomel electrode (SCE) as reference in a Luggin side arm. In this article, all potential values E have been adjusted to the standard hydrogen electrode (SHE) scale by using the relationship $E_{SHE} = E_{SCE} + 242$ mV. The cell was blacked out with black electrical tape to avoid light-activated processes convoluting the results (7, 16) and thermostated by a water-jacket linked to a water-circulator. To prepare an enzyme film, the PGE electrode was first polished for 10 s with an aqueous slurry of α-alumina (1 μm, Buehler) and sonicated for 5 s in purified water, before 1.5 μL enzyme solution (containing 0.02–0.2 μg CrHydA1, i.e., up to ~2 μM, pH 8) was applied and removed after a few minutes.

Electrochemical experiments were performed by using an electrochemical analyzer (Autolab PGSTAT20) controlled by a computer operating GPES software (EcoChemie). Mass flow controllers (Smart-Trak Series 100, Sierra Instruments) were used to prepare precise gas mixtures (accurate to within 1%) and to impose constant gas flow rates into the electrochemical cell during experiments. Depending on the nature of the experiment and the rate of the reaction, changes of gas conditions were performed by (i) changing the gas

RESULTS

mixture reaching the cell headspace, or (ii) by a simultaneous change in gas mixture in the headspace and injection of an aliquot of solution equilibrated with that gas. Efficient mixing and gas-solution equilibration were achieved through rapid electrode rotation (3,000 rpm).

X-Ray Absorption Measurements. K_α-fluorescence-detected XAS spectra at the Fe K-edge were collected at T = 20 K using an energy-resolving 13-element Ge detector and a helium cryostat as previously described (28, 29) at beamline D2 of the EMBL outstation (at HASYLAB, DESY). Harmonic rejection was achieved by detuning of the Si (111) double-crystal monochromator to 50% of its peak intensity. Spectra were collected maximally for a scan range of 6,950–8,450 eV. Deadtime-corrected XAS spectra were averaged after energy calibration of each scan by using the peak at 7,112 eV in the 1st derivative of the absorption spectrum of an Fe-foil as an energy standard (estimated accuracy ± 0.1 eV) (28, 29). Data were then normalized, and extended EXAFS oscillations were extracted (39). The energy scale of EXAFS spectra was converted to the wavevector scale (k-scale) by using an E_0 value of 7,112 eV. Unfiltered k^3-weighted spectra were used for least-squares curve-fitting employing a multiple-scattering approach with the program EXCURV (40). Fourier transforms were calculated from k^3-weighted EXAFS data by using the program SimX (39) and employing $\cos^2$ windows ranging >10% at both ends of the k-range. From experimental K-edge spectra the preedge peak region was extracted by using the program Xanda (www.bit.ly/1V1zKe).

ACKNOWLEDGMENTS. We thank the beamline scientist Dr. W. Meyer Klaucke (EMBL at DESY) for excellent technical support. We acknowledge Professor Achim Trebst for his teachings and for ongoing insightful discussions that made this work possible. We dedicate this article to him on the occasion of his 80th birthday. This work was funded by the Deutsche Forschungsgemeinschaft Grants SFB480 (to T.H.) and SFB498-C8 (to M.H.), the "Unicat" Cluster of Excellence Berlin, the EU/Energy Network SolarH2 (FP7 contract 212508), and Biotechnology and Biological Sciences Research Council Grant BB/D52222X/1 (to F.A.A.). K.A.V. is a Royal Society Research Fellow.

1. Vignais PM, Billoud B (2007) Occurrence, classification, and biological function of hydrogenases: An overview. *Chem Rev* 107:4206–4272.
2. Cammack R, Frey M, Robson R (2001) in *Hydrogen As a Fuel: Learning from Nature* (Taylor and Francis, London).
3. Adams MWW (1990) The structure and mechanism of iron hydrogenases. *Biochim Biophys Acta* 1020:115–145.
4. Parkin A, Goldet G, Cavazza C, Fontecilla-Camps JC, Armstrong FA (2008) The difference a Se makes? Oxygen-tolerant hydrogen production by the [NiFeSe]-hydrogenase from *Desulfomicrobium baculatum*. *J Am Chem Soc* 130:13410–13416.
5. Vincent KA, Parkin A, Armstrong FA (2007) Investigating and exploiting the electrocatalytic properties of hydrogenases. *Chem Rev* 107:4366–4413.
6. Hambourger M, et al. (2008) [FeFe]-hydrogenase-catalyzed H_2 production in a photoelectrochemical biofuel cell. *J Am Chem Soc* 130:2015–2022.
7. Parkin A, Cavazza C, Fontecilla-Camps JC, Armstrong FA (2006) Electrochemical investigations of the interconversions between catalytic and inhibited states of the [FeFe]-hydrogenase from *Desulfovibrio desulfuricans*. *J Am Chem Soc* 128:16808–16815.
8. Ghirardi ML, Dubini A, Yu JP, Maness PC (2009) Photobiological hydrogen-producing systems. *Chem Soc Rev* 38:52–61.
9. Nicolet Y, Piras C, Legrand P, Hatchikian C, Fontecilla-Camps JC (1999) *Desulfovibrio desulfuricans* iron hydrogenase: The structure shows unusual coordination to an active site Fe binuclear center. *Structure* 7:13–23.
10. Tard C, et al. (2005) Synthesis of the H-cluster framework of iron-only hydrogenase. *Nature* 433:610–613.
11. Peters JW, Lanzilotta WN, Lemon BJ, Seefeldt LC (1998) X-ray crystal structure of the Fe-only hydrogenase (CpI) from *Clostridium pasteurianum* to 1.8 angstrom resolution. *Science* 282:1853–1858.
12. Pandey AS, Harris TV, Giles LJ, Peters JW, Szilagyi RK (2008) Dithiomethylether as a ligand in the hydrogenase H-cluster. *J Am Chem Soc* 130:4533–4540.
13. Nicolet Y, De Lacey AL, Vernede X, Fernandez VM, Hatchikian EC, Fontecilla-Camps JC (2001) Crystallographic and FTIR spectroscopic evidence of changes in Fe coordination upon reduction of the active site of the Fe-only hydrogenase from *Desulfovibrio desulfuricans*. *J Am Chem Soc* 123:1596–1601.
14. Silakov A, Reijerse EJ, Albracht SPJ, Hatchikian EC, Lubitz W (2007) The electronic structure of the H-cluster in the [FeFe]-hydrogenase from *Desulfovibrio desulfuricans*: A Q-band Fe-57-ENDOR and HYSCORE study. *J Am Chem Soc* 129:11447–11458.
15. Lubitz W, Reijerse E, van Gastel M (2007) [NiFe] and [FeFe] hydrogenases studied by advanced magnetic resonance techniques. *Chem Rev* 107:4331–4365.
16. Roseboom W, Lacey AL, Fernandez VM, Hatchikian EC, Albracht SPJ (2006) The active site of the [FeFe]-hydrogenase from *Desulfovibrio desulfuricans*. II. Redox properties, light sensitivity, and CO-ligand exchange as observed by infrared spectroscopy. *J Biol Inorg Chem* 11:102–118.
17. Lemon BJ, Peters JW (1999) Binding of exogenously added carbon monoxide at the active site of the iron-only hydrogenase (CpI) from *Clostridium pasteurianum*. *Biochemistry* 38:12969–12973.
18. De Lacey AL, Stadler C, Cavazza C, Hatchikian C, Fernandez VM (2000) FTIR characterization of the active site of the Fe-hydrogenase from *Desulfovibrio desulfuricans*. *J Am Chem Soc* 122:11232–11233.
19. Cotton FA (1988) in *Advanced Inorganic Chemistry*, ed Cotton FA (Wiley, New York), p 58.
20. Kubas GJ (2007) Fundamentals of H_2 binding and reactivity on transition metals underlying hydrogenase function and H_2 production and storage. *Chem Rev* 107:4152–4205.
21. Erbes DL, King D, Gibbs M (1979) Inactivation of hydrogenase in cell-free-extracts and whole cells of *Chlamydomonas reinhardtii* by oxygen. *Plant Physiol* 63:1138–1142.
22. Crack JC, Green J, Cheesman MR, Le Brun NE, Thomson AJ (2007) Superoxide-mediated amplification of the oxygen-induced switch from [4Fe-4S] to [2Fe-2S] clusters in the transcriptional regulator FNR. *Proc Natl Acad Sci USA* 104:2092–2097.
23. Imlay JA (2006) Iron-sulphur clusters and the problem with oxygen. *Mol Microbiol* 59:1073–1082.
24. Kamp C, Silakov A, Winkler M, Reijerse EJ, Lubitz W, Happe T (2008) Isolation and first EPR characterization of the [FeFe]-hydrogenases from green algae. *Biochim Biophys Acta* 1777:410–416.
25. Stripp S, Sanganas O, Happe T, Haumann M (2009) The structure of the active site H-cluster of [FeFe] hydrogenase from the green alga *Chlamydomonas reinhardtii* studied by X-ray absorption spectroscopy. *Biochemistry* 48:5042–5049.
26. Winkler M, Heil B, Heil B, Happe T (2002) Isolation and molecular characterization of the [Fe]-hydrogenase from the unicellular green alga. *Chlorella fusca Biochim Biophys Acta* 1576:330–334.
27. Baffert C, et al. (2008) Hydrogen-activating enzymes: Activity does not correlate with oxygen sensitivity. *Angew Chem-Int Ed* 47:2052–2054.
28. Buhrke T, et al. (2005) Reduction of unusual iron-sulfur clusters in the H_2-sensing regulatory Ni-Fe hydrogenase from *Ralstonia eutropha* H16. *J Biol Chem* 280:19488–19495.
29. Loescher S, Schwartz L, Stein M, Ott S, Haumann M (2007) Facilitated hydride binding in an Fe-Fe hydrogenase active-site biomimic revealed by X-ray absorption spectroscopy and DFT calculations. *Inorg Chem* 46:11094–11105.
30. Goldet G, et al. (2009) Dynamic electrochemical investigations of hydrogen oxidation and production by enzymes and implications for future technology. *Chem Soc Rev* 38:36–51.
31. Fontecilla-Camps JC, Volbeda A, Cavazza C, Nicolet Y (2007) Structure/function relationships of [NiFe]- and [FeFe]-hydrogenases. *Chem Rev* 107:4273–4303.
32. Messner KR, Imlay JA (2002) Mechanism of superoxide and hydrogen peroxide formation by fumarate reductase, succinate dehydrogenase, and aspartate oxidase. *J Biol Chem* 277:42563–42571.
33. Varghese S, Tang Y, Imlay JA (2003) Contrasting sensitivities of *Escherichia coli* aconitases A and B to oxidation and iron depletion. *J Bacteriol* 185:221–230.
34. Tilley GJ, Camba R, Burgess BK, Armstrong FA (2001) Influence of electrochemical properties in determining the sensitivity of [4Fe-4S] clusters in proteins to oxidative damage. *Biochem J* 360:717–726.
35. Camba R, Armstrong FA (2000) Investigations of the oxidative disassembly of Fe-S clusters in *Clostridium pasteurianum* 8Fe ferredoxin using pulsed-protein-film voltammetry. *Biochemistry* 39:10587–10598.
36. Crack JC, Le Brun NE, Thomson AJ, Green J, Jervis AJ (2008) in *Globins and Other Nitric Oxide-Reactive Proteins, Part B* (Elsevier, San Diego, CA), pp 171–209.
37. Goldet G, et al. (2008) Hydrogen production under aerobic conditions by membrane-bound hydrogenases from *Ralstonia species*. *J Am Chem Soc* 130:11106–11113.
38. von Abendroth G, et al. (2008) Optimized over-expression of [FeFe] hydrogenases with high specific activity in *Clostridium acetobutylicum*. *Int J Hyd En* 33:6076–6081.
39. Dau H, Liebisch P, Haumann M (2003) X-ray absorption spectroscopy to analyze nuclear geometry and electronic structure of biological metal centers - Potential and questions examined with special focus on the tetra-nuclear manganese complex of oxygenic photosynthesis. *Anal Bioanal Chem* 376:562–583.
40. Tomic S, et al. (2004) New Tools for the Analysis of EXAFS: The DL-EXCURV Package. *CCLRC Technical Report* 1–10.

RESULTS

2.5 Electrochemical Kinetic Investigations of the Reactions of [FeFe]-hydrogenases with Carbon Monoxide and Oxygen: Comparing the Importance of Gas Tunnels and Active-Site Electronic/Redox Effects

Gabrielle Goldet,[1] Caterina Brandmayr,[1] **Sven T. Stripp**,[2] Thomas Happe,[2] Christine Cavazza,[3] Juan C. Fontecilla-Camps[3] and Fraser A. Armstrong*[1]

[1]Inorganic Chemistry Laboratory, Department of Chemistry, University of Oxford, South Parks Road, Oxford OX1 3QR, UK;

[2]Ruhr-Universitat, Lehrstuhl für Biochemie der Pflanzen, AG Photobiotechnologie, 44780 Bochum, Germany;

[3]Laboratoire de Cristallographie et Cristallographie des Protéines, Institut de Biologie Structurale, J.P. Ebel, CEA, CNRS, Université Joseph Fourier, 41, rue J. Horrowitz, 38027 Grenoble Cedex 1, France

* Corresponding Author:
Phone: + 44 (0) 1865 272647
E-mail: fraser.armstrong@chem.ox.ac.uk

http://dx.doi.org/ 10.1021/ja905388j

RESULTS

Electrochemical Kinetic Investigations of the Reactions of [FeFe]-Hydrogenases with Carbon Monoxide and Oxygen: Comparing the Importance of Gas Tunnels and Active-Site Electronic/Redox Effects

Gabrielle Goldet,[†] Caterina Brandmayr,[†] Sven T. Stripp,[‡] Thomas Happe,[‡] Christine Cavazza,[§] Juan C. Fontecilla-Camps,[§] and Fraser A. Armstrong*,[†]

Inorganic Chemistry Laboratory, Department of Chemistry, University of Oxford, South Parks Road, Oxford OX1 3QR, United Kingdom, Ruhr-Universität, Lehrstuhl für Biochemie der Pflanzen, AG Photobiotechnologie, 44780 Bochum, Germany, and Laboratoire de Cristallographie et Cristallographie des Protéines, Institut de Biologie Structurale, J.P. Ebel, CEA, CNRS, Université Joseph Fourier, 41, rue J. Horrowitz, 38027 Grenoble Cedex 1, France

Received July 3, 2009; E-mail: fraser.armstrong@chem.ox.ac.uk

Abstract: A major obstacle for future biohydrogen production is the oxygen sensitivity of [FeFe]-hydrogenases, the highly active catalysts produced by bacteria and green algae. The reactions of three representative [FeFe]-hydrogenases with O_2 have been studied by protein film electrochemistry under conditions of both H_2 oxidation and H_2 production, using CO as a complementary probe. The hydrogenases are DdHydAB and CaHydA from the bacteria *Desulfovibrio desulfuricans* and *Clostridium acetobutylicum*, and CrHydA1 from the green alga *Chlamydomonas reinhardtii*. Rates of inactivation depend on the redox state of the active site 'H-cluster' and on transport through the protein to reach the pocket in which the H-cluster is housed. In all cases CO reacts much faster than O_2. In the model proposed, CaHydA shows the most sluggish gas transport and hence little dependence of inactivation rate on H-cluster state, whereas DdHydAB shows a large dependence on H-cluster state and the least effective barrier to gas transport. All three enzymes show a similar rate of reactivation from CO inhibition, which increases upon illumination: the rate-determining step is thus assigned to cleavage of the labile Fe-CO bond, a reaction likely to be intrinsic to the atomic and electronic state of the H-cluster and less sensitive to the surrounding protein.

Introduction

The increasing need for clean, renewable fuels is stimulating new research on hydrogen (H_2) production,[1−5] and one promising solution is to exploit microorganisms in 'H_2 farms'. In biology, H_2 is evolved by metalloenzymes called hydrogenases, in processes ranging from fermentation to photosynthesis. Hydrogenases are highly efficient enzymes—so much so that when attached to an electrode, they are, like platinum, superb electrocatalysts of both H_2 oxidation and H_2 production, at or close to the reversible potential for the $2H^+/H_2$ couple.[6−8]

Of the two main classes, [NiFe]- and [FeFe]-hydrogenases, named according to the metals present in the center at which H_2 is activated, the [FeFe]-hydrogenases are considered to be more active in H_2 production.[9] However, a perceived major disadvantage of [FeFe]-hydrogenases (with respect to [NiFe]-hydrogenases) is their higher O_2 sensitivity.[10] The [NiFe]-hydrogenases react rapidly with O_2 to give inactive, EPR-characterized, Ni(III) forms that can be reactivated by reduction: Ni-A ('unready') is reactivated very slowly whereas Ni-B ('ready') can be reactivated within seconds, hence there is a rapid repair mechanism for hydrogenases that produce only Ni-B.[6,11] In contrast, the [FeFe]-hydrogenases appear to undergo irreparable damage when exposed to O_2 while in their active state (after reduction).[12−14] The incompatibility of O_2 with [FeFe]-hydrogenases poses a major limitation to progress in 'biohydrogen' production, in particular by modified photosyn-

---

[†] University of Oxford.
[‡] Ruhr-Universität.
[§] Université Joseph Fourier.
(1) Navarro, R. M.; Pena, M. A.; Fierro, J. L. G. *Chem. Rev.* **2007**, *107*, 3952−3991.
(2) Mertens, R.; Liese, A. *Curr. Opin. Biotechnol.* **2004**, *15*, 343−348.
(3) Hankamer, B.; Lehr, F.; Rupprecht, J.; Mussgnug, J. H.; Posten, C.; Kruse, O. *Physiol. Plant.* **2007**, *131*, 10−21.
(4) Liu, X. M.; Ibrahim, S. K.; Tard, C.; Pickett, C. J. *Coord. Chem. Rev.* **2005**, *249*, 1641−1652.
(5) Melis, A.; Happe, T. *Plant Physiol.* **2001**, *127*, 740−748.
(6) Vincent, K. A.; Parkin, A.; Armstrong, F. A. *Chem. Rev.* **2007**, *107*, 4366−4413.
(7) Jones, A. K.; Sillery, E.; Albracht, S. P. J.; Armstrong, F. A. *Chem. Commun.* **2002**, 866−867.
(8) Hambourger, M.; Gervaldo, M.; Svedruzic, D.; King, P. W.; Gust, D.; Ghirardi, M.; Moore, A. L.; Moore, T. A. *J. Am. Chem. Soc.* **2008**, *130*, 2015−2022.

(9) Frey, M. *ChemBioChem* **2002**, *3*, 153−160.
(10) Adams, M. W. W. *Biochim. Biophys. Acta* **1990**, *1020*, 115−145.
(11) Armstrong, F. A.; Belsey, N. A.; Cracknell, J. A.; Goldet, G.; Parkin, A.; Reisner, E.; Vincent, K. A.; Wait, A. F. *Chem. Soc. Rev.* **2009**, *38*, 36−51.
(12) Erbes, D. L.; King, D.; Gibbs, M. *Plant Physiol.* **1979**, *63*, 1138−1142.
(13) Erbes, D. L.; King, D.; Gibbs, M. *Plant Physiol.* **1978**, *61*, 23−23.
(14) Vincent, K. A.; Parkin, A.; Lenz, O.; Albracht, S. P. J.; Fontecilla-Camps, J. C.; Cammack, R.; Friedrich, B.; Armstrong, F. A. *J. Am. Chem. Soc.* **2005**, *127*, 18179−18189.

45

RESULTS

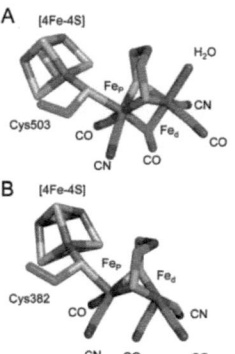

Figure 1. Structures of H-clusters of the [FeFe]-hydrogenase from *Clostridium pasteurianum* (*Cp*I) and *Desulfovibrio desulfuricans* (*Dd*HydAB) constructed using PyMol. A) *Cp*I H-cluster (PDB code: 3C8Y).[23] B) *Dd*HydAB H-cluster.[25] The two structures were modeled with different bridgehead atoms—O for *Cp*I, and N for *Dd*HydAB—but this distinction is not directly relevant for this paper.

thesis.[15] In green algae the O_2-sensitivity of the [FeFe]-hydrogenase is the bottleneck for producing H_2 from sunlight. Production of H_2 is stimulated during sulfur deprivation, conditions under which only 10% of the photosystem II remains active and the system effectively becomes anaerobic.[16] Green algae that could express an O_2-tolerant [FeFe]-hydrogenase would therefore provide much increased levels of H_2 production.[15]

The buried active site of [FeFe]-hydrogenases is actually a complex 6Fe unit known as the 'H-cluster' which contains a [4Fe-4S] subcluster (generally referred to as [4Fe-4S]$_H$) in addition to the di-iron subcluster (2Fe$_H$).[17] The two independent representations of the H-cluster shown in Figure 1 are directly relevant to the catalytically active states known as H_{ox} and H_{red} that have been extensively characterized.[18,19] General features of the structure are as follows: (a) the [4Fe-4S]$_H$ subcluster is linked to one of the Fe atoms of the 2Fe$_H$ subcluster by a bridging cysteine sulfur (the Fe atoms of 2Fe$_H$ are thus known as 'proximal' (Fe$_p$) and 'distal' (Fe$_d$) with respect to the [4Fe-4S]$_H$ subcluster);[20] (b) both Fe$_d$ and Fe$_p$ are coordinated by CO and CN$^-$ ligands;[17] (c) an unusual SCH$_2$XCH$_2$S dithiolate ligand forms a di-μ-thiolato bridge between Fe$_p$ and Fe$_d$, and although opinions differ as to whether the bridgehead atom X is an O or N atom,[21-25] recent investigations with ^{14}N HYSCORE have provided direct evidence that X = N.[21]

In the structure of H_{ox}, as determined from the *Cp*I enzyme from *Clostridium pasteurianum*, Fe$_p$ is coordinated by one CO and one CN$^-$ ligand and shares a bridging CO with Fe$_d$.[20] In turn, Fe$_d$ is also coordinated by one CO and one CN$^-$ ligand, and an additional binding site is vacant or occupied by an exchangeable O-ligand, most likely a water molecule (Figure 1A). In the structure of the [FeFe]-hydrogenase from *Desulfovibrio desulfuricans*, which should be in the H_{red} form, the bridging CO is replaced by a terminal CO on Fe$_d$ (Figure 1B).[26] Recent EPR spectroscopic investigations favor an oxidation state assignment of [4Fe-4S]$^{2+}$-Fe$_p$(I)Fe$_d$(II) for H_{ox}, with some spin density delocalized onto the [4Fe-4S]$_H$ subcluster,[27] although H-clusters from different enzymes show minor variations in electronic structure.[28] The EPR-silent H_{red} state is assigned as [4Fe-4S]$^{2+}$-Fe(I)Fe(I) which, if protonated, is formally equivalent to the hydrido species [4Fe-4S]$^{2+}$-Fe(II)Fe(II)-H$^-$.[17] As also determined by EPR spectroscopy, exogenous CO, a competitive inhibitor of H_2 oxidation, reacts with H_{ox}.[29] Crystallographic and infrared spectroscopic studies of H_{ox}-CO further show that binding of CO (which is photolabile) occurs at Fe$_d$.[30-33] Inactivation by anaerobic oxidants gives rise to a form known as H_{ox}^{inact}, usually formulated as [4Fe-4S]$^{2+}$-Fe(II)Fe(II), which can be reactivated upon reduction—a process occurring via an intermediate known as H_{trans} which has been formulated as [4Fe-4S]$^+$-Fe(II)Fe(II), i.e. with the [4Fe-4S]$_H$ subcluster reduced.[33] The H-cluster is remarkable among non-macrocycle cofactors because the 2Fe$_H$ subcluster at which H_2 is produced is connected to the protein through just a half-share of a cysteine sulfur: it is very much an organometallic-like compound physically enclosed in protein.

Despite these intense studies by crystallography and spectroscopy, numerous aspects of the reactions of [FeFe]-hydrogenases remain unresolved. These aspects include the activation process (there is evidence from electrochemical titrations that a two-electron process is also involved)[33] and many details of the mechanism of catalysis in either direction, including the exact function of the [4Fe-4S]$_H$ subcluster. The nature and products of the degradation by O_2 are only now coming to light,[34] and a major issue is whether and how H_2 production could be sustainable at all in the presence of O_2.

This article describes mechanistic investigations, by protein film electrochemistry, of the O_2 inactivation kinetics of three representative [FeFe]-hydrogenases. These are: the hydrogenase

(15) Ghirardi, M. L.; Posewitz, M. C.; Maness, P.-C.; Dubini, A.; Yu, J.; Seibert, M. *Annu. Rev. Plant Biol.* **2007**, *58*, 71–91.
(16) Hemschemeier, A.; Fouchard, S.; Cournac, L.; Peltier, G.; Happe, T. *Planta* **2008**, *227*, 397–407.
(17) Fontecilla-Camps, J. C.; Volbeda, A.; Cavazza, C.; Nicolet, Y. *Chem. Rev.* **2007**, *107*, 4273–4303.
(18) Lubitz, W.; Reijerse, E.; van Gastel, M. *Chem. Rev.* **2007**, *107*, 4331–4365.
(19) De Lacey, A. L.; Fernandez, V. M.; Rousset, M.; Cammack, R. *Chem. Rev.* **2007**, *107*, 4304–4330.
(20) Nicolet, Y.; Lemon, B. J.; Fontecilla-Camps, J. C.; Peters, J. W. *Trends Biochem. Sci.* **2000**, *25*, 138–143.
(21) Silakov, A.; Wenk, B.; Reijerse, E.; Lubitz, W. *Phys. Chem. Chem. Phys.* **2009**, *11*, 6592–9.

(22) Peters, J. W.; Lanzilotta, W. N.; Lemon, B. J.; Seefeldt, L. C. *Science* **1998**, *282*, 1853–1858.
(23) Pandey, A. S.; Harris, T. V.; Giles, L. J.; Peters, J. W.; Szilagyi, R. K. *J. Am. Chem. Soc.* **2008**, *130*, 4533–4540.
(24) Barton, B. E.; Olsen, M. T.; Rauchfuss, T. B. *J. Am. Chem. Soc.* **2008**, *130*, 16834–5.
(25) Nicolet, Y.; De Lacey, A. L.; Vernede, X.; Fernandez, V. M.; Hatchikian, E. C.; Fontecilla-Camps, J. C. *J. Am. Chem. Soc.* **2001**, *123*, 1596–1601.
(26) Nicolet, Y.; Piras, C.; Legrand, P.; Hatchikian, C.; Fontecilla-Camps, J. C. *Structure* **1999**, *7*, 13–23.
(27) Silakov, A.; Reijerse, E. J.; Albracht, S. P. J.; Hatchikian, E. C.; Lubitz, W. *J. Am. Chem. Soc.* **2007**, *129*, 11447–11458.
(28) Kamp, C.; Silakov, A.; Winkler, M.; Reijerse, E. J.; Lubitz, W.; Happe, T. *Biochim. Biophys. Acta* **2008**, *1777*, 410–416.
(29) Bennett, B.; Lemon, B. J.; Peters, J. W. *Biochemistry* **2000**, *39*, 7455–7460.
(30) Lemon, B. J.; Peters, J. W. *J. Am. Chem. Soc.* **2000**, *122*, 3793–3794.
(31) Lemon, B. J.; Peters, J. W. *Biochemistry* **1999**, *38*, 12969–12973.
(32) De Lacey, A. L.; Stadler, C.; Cavazza, C.; Hatchikian, E. C.; Fernandez, V. M. *J. Am. Chem. Soc.* **2000**, *122*, 11232–11233.
(33) Roseboom, W.; Lacey, A. L.; Fernandez, V. M.; Hatchikian, E. C.; Albracht, S. P. J. *J. Biol. Inorg. Chem.* **2006**, *11*, 102–118.
(34) Stripp, S.; Goldet, G.; C., B.; Vincent, K. A.; Armstrong, F. A.; Happe, T. *Proc. Natl. Acad. Sci. U.S.A.* **2009**, in press.

from a sulfate-reducing bacterium, *Desulfovibrio desulfuricans*, abbreviated as *Dd*HydAB which has been crystallographically characterized;[26] the hydrogenase from *Clostridium acetobutylicum*, abbreviated as *Ca*HydA, which is potentially of importance for H_2 production by anaerobic fermentation and has high sequence similarity with the crystallographically characterized *Cp*I hydrogenase from *C. pasteurianum*,[22] and the hydrogenase known as *Cr*HydA1 from the green alga *Chlamydomonas reinhardtii*, which is of interest for photosynthetic H_2 production.[35] Both *Dd*HydAB and *Ca*HydA contain a series of Fe−S clusters[17] to relay electrons within the protein for transfer to and from the redox partner (in our case the electrode); in contrast *Cr*HydA1 possesses no Fe−S clusters apart from the [4Fe-4S]$_H$ subcluster,[36,37] but it is nonetheless electrocatalytically active when adsorbed on an electrode.[34] Although the three enzymes differ in their overall tertiary and quaternary structures, their H-domains that house the H-cluster are very similar.[17]

In protein film electrochemistry, an enzyme is immobilized on the surface of an electrode such that its properties are controlled directly by the electrode potential.[6,38] Catalytic activity in either direction, oxidation or reduction, can be driven and recorded at any particular potential value and the catalytic rate is directly proportional to the current that flows. Various gas mixtures (produced by mass-flow controllers) can be introduced and flushed from the sealed electrochemical cell in which the electrode is rotated rapidly to provide precise hydrodynamic control (supply of reactants and removal of products). A particular advantage of this approach is that turnover activity in either direction is immediately and directly observed from the catalytic current; thus, *rates of change of activity*, such as those induced by CO or O_2, are extracted directly from the variation of current with time, all as a precise function of the electrode potential. In this way, extremely complex reactivities become resolvable; therefore, this technique both complements and instigates structural and spectroscopic investigations.

We first establish, for each enzyme, how H_2 oxidation and H_2 production are affected by the concentrations of H_2 and CO; next we examine the kinetics of CO binding and release and correlate these data with equilibrium values; we then examine the kinetics of O_2 inactivation of H_2 oxidation activity; finally, we exploit CO inhibition as a tool to investigate H_2 production in the presence of O_2. Recent studies have established that CO is able to protect [FeFe]-hydrogenases against inactivation by O_2,[34,39] an observation suggesting that the sequence of destruction is initiated by O_2 coordinating to the same site at which exogenous CO binds, i.e. Fe$_d$.[30−33] Thereafter, the mechanism remains less clear, but recent EXAFS evidence obtained with *Cr*HydA1 shows that the [4Fe-4S]$_H$ subcluster is altered more than the 2Fe$_H$ subcluster.[34] Our experiments show clearly how the destructive power of O_2 varies among the hydrogenases with an interesting dependence on catalytic direction (H_2 oxidation compared to H_2 production)—thus implicating sensitivity to the oxidation level of the active site. The results provide insight for the quest for solutions to the oxygen problem in biohydrogen production by photosynthesis. The functional differences between the [FeFe]-hydrogenases are not immediately evident from the structures that have been obtained for *Dd*HydAB and *Cp*I.

Methods

Previously reported protocols were employed to obtain pure samples of *Dd*HydAB,[40] *Cr*HydA1 and *Ca*HydA.[41] In each case levels of CO, O_2 and H_2 present during cell growth were extremely low. Protein film electrochemistry experiments were carried out in an anaerobic glovebox (M Braun) comprising a N_2 atmosphere. The solutions contained 0.05 M phosphate buffer with 0.10 M NaCl as additional supporting electrolyte, and were prepared using standard reagents NaCl, NaH$_2$PO$_4$ and Na$_2$HPO$_4$ (Analytical Reagent grade, Sigma) in purified water (Millipore 18 MΩ cm). The working electrode was a disk (area 0.03 cm^2) of pyrolytic graphite oriented so that the 'edge plane' faced the solution. This is called a pyrolytic graphite edge (PGE) electrode and it was used in conjunction with an electrode rotator (EcoChemie) that fitted snugly into a specially designed, gastight, glass electrochemical cell. The cell featured a water jacket for temperature control. Due to the light sensitivity of [FeFe]-hydrogenases the entire surface of the cell was blacked-out with adhesive masking tape. In experiments to detect photolabilization, the tape at the bottom of the cell was removed to allow illumination from a 150 W lamp placed just beneath the cell (see Figure SI.1, Supporting Information). A saturated calomel reference electrode (SCE) was placed in a side arm containing 0.1 M NaCl, connected to the main cell compartment by a Luggin capillary. A Pt wire was used as the counter electrode. Potentials (E) are quoted with respect to the standard hydrogen electrode (SHE) using the correction $E_{SHE} = E_{SCE} + 242$ mV at 298 K.[42] Electrochemical experiments were performed using an electrochemical analyzer (Autolab PGSTAT10 or 20) controlled by a computer operating GPES software (EcoChemie). Mass flow controllers (Smart-Trak Series 100, Sierra Instruments, U.S.A.) were used to prepare precise gas mixtures (headspace fractions accurate to within 1%) and to impose constant gas flow rates into the electrochemical cell during experiments. Gases used were H_2 (Premier grade, Air Products), O_2 (Air Products), CO (Research grade, BOC), 1% CO in N_2 (Research grade, BOC), N_2 (Oxygen-free, BOC) or mixtures of these gases. Henry's law was applied to estimate the concentration of gas in solution in each experiment.[43] Values taken for use at 10 °C were: 100% CO, 1248 μM; 100% O_2, 1624 μM.

To prepare each enzyme film, the PGE electrode was first polished for 10 s with an aqueous slurry of α-alumina (1 μm, Buehler) and sonicated for 5 s in purified water, before enzyme solution (1.5 μL, containing 0.015−0.15 μg, pH 8) was applied and removed after a few minutes. The electrode was then placed in enzyme-free buffered electrolyte so that the only enzyme molecules being addressed were on the electrode and subjected to the same regime of strict potential control. In all experiments the electrode was rotated at a constant high rate (3000−5000 rpm) to ensure efficient supply of substrate and removal of product.

In experiments measuring the effect of O_2 on hydrogenase-catalyzed H_2 production, it was necessary to estimate the concentration of O_2 that the enzyme molecules at the electrode surface actually experience, given that a portion of the total dissolved O_2 is consumed by the electrode at this potential (−0.4 V). Using a

---

(35) Ghirardi, M. L.; Dubini, A.; Yu, J. P.; Maness, P. C. *Chem. Soc. Rev.* **2009**, *38*, 52−61.
(36) Stripp, S.; Sanganas, O.; Happe, T.; Haumann, M. *Biochemistry* **2009**, *48*, 5042−5049.
(37) Happe, T.; Naber, J. D. *Eur. J. Biochem.* **1993**, *214*, 475−481.
(38) Léger, C.; Bertrand, P. *Chem. Rev.* **2008**, *108*, 2379−2438.
(39) Baffert, C.; Demuez, M.; Cournac, L.; Burlat, B.; Guigliarelli, B.; Bertrand, P.; Girbal, L.; Leger, C. *Angew. Chem., Int. Ed.* **2008**, *47*, 2052−2054.

(40) Hatchikian, C.; Forget, N.; Fernandez, V. M.; Williams, R.; Cammack, R. *Eur. J. Biochem.* **1992**, *209*, 357−365.
(41) von Abendroth, G.; Stripp, S.; Silakov, A.; Croux, C.; Soucaille, P.; Girbal, L.; Happe, T. *Int. J. Hydrogen Energy* **2008**, *33*, 6076−6081.
(42) Bard, A. J.; Faulkner, L. R. *Electrochemical Methods. Fundamentals and Applications*, 2nd ed.; Wiley: New York, 2001.
(43) Sander, R. http://www.henrys-law.org, 1999.

RESULTS

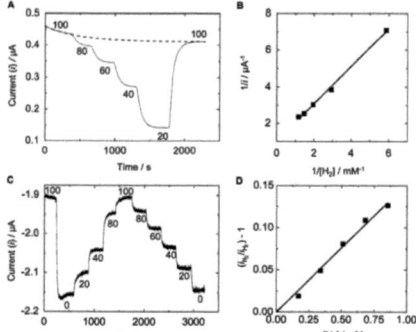

Figure 2. Determination of $K_M^{H_2}$ (A and B) and K_I^{app} (C and D) for CaHydA. (Panel A) chronoamperometric experiment performed at −0.05 V, at pH 6, 10 °C. The broken line traces the progression of film loss throughout the experiment. (Panel B) Lineweaver−Burk plot from experiment shown in A fitted to a straight line. (Panel C) chronoamperometric experiment performed to determine K_I^{app} at −0.4 V, at pH 6, 10 °C. The concentration of H_2 in N_2 (%) in the headspace of the cell is indicated. The rotation rate was varied from 3000 to 4000 and 5000 rpm at each concentration of H_2 to ensure that inhibition was not mass-transport limited. (Panel D) Plot according to the procedure described by Léger et al.[45] from which K_I^{app} is determined, showing the line of best fit.

previously published method,[44] it was estimated that 0.7% O_2 (11 μM in solution at 10 °C) survives to be experienced by the enzyme when 1% O_2 is flowed through the headspace of the cell and 4.5% (73 μM) survives when 5% O_2 is used.

Results

Measurements of the H_2 Concentration Dependencies for H_2 Oxidation and Production. We first evaluated the affinity of the enzymes for H_2 both in terms of K_M for H_2 as the substrate in H_2 oxidation ($K_M^{H_2}$) and the apparent inhibition constant K_I^{app} for H_2 as the product inhibitor of H^+ reduction. Figure 2 shows experiments carried out for CaHydA: analogous experiments were carried out for CrHydA1, but films of DdHydAB were not sufficiently stable to obtain accurate measurements over the period of time required (30−60 min.). Experiments to determine $K_M^{H_2}$ were conducted by varying the ratio of H_2 to N_2 in the headgas and measuring the oxidation current after allowing time for the gas mixture to equilibrate with the cell solution at each concentration of H_2 (Figure 2A). The experiments were performed at an electrode potential of −0.05 V to avoid the anaerobic inactivation that occurs at higher potential (see Figure 3). We and others, have noted that K_M and K_I are potential-dependent quantities,[44,45] and the potential must therefore be specified. It was also important to make measurements under conditions where the current was limited by the catalytic rate of the enzyme rather than by mass transport of substrate to (or product from) the electrode. The experiments were carried out at low temperature (10 °C) to decrease the rate of catalysis. The low temperature also minimized the

(44) Goldet, G.; Wait, A. F.; Cracknell, J. A.; Vincent, K. A.; Ludwig, M.; Lenz, O.; Friedrich, B.; Armstrong, F. A. *J. Am. Chem. Soc.* **2008**, *130*, 11106−11113.
(45) Léger, C.; Dementin, S.; Bertrand, P.; Rousset, M.; Guigliarelli, B. *J. Am. Chem. Soc.* **2004**, *126*, 12162−12172.

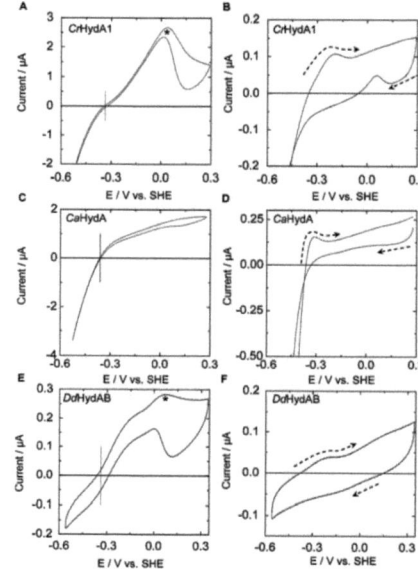

Figure 3. Cyclic voltammograms showing bidirectional electrocatalytic H^+ reduction and H_2 oxidation by CrHydA1 (A), CaHydA (C) and DdHydAB (E) at pH 6.0 under 1 bar H_2. Panels B, D, F show the time-dependent voltammograms recorded following introduction of CO, introduced prior to the scans shown here, as the potential was being cycled between −0.55 V and +0.3 V, as it is removed from the cell: these reveal different reactions as CO dissociates and rebinds as a function of potential. Experimental conditions for CrHydA1: 10 °C, scan rate 20 mV/s, inhibition was achieved by injection of CO-saturated buffer to give an immediate concentration of 100 μM CO in solution 5 min prior to the scan shown in panel B. Experimental conditions for CaHydA: 32 °C, electrode rotation 3000 rpm, scan rate 20 mV/s, inhibition was achieved by flushing 100% CO through the cell 5 min prior to the scan shown in panel D. Experimental conditions for DdHydAB: 10 °C, electrode rotation 3000 rpm, scan rate 10 mV/s, inhibition was achieved by injection of CO-saturated buffer to give an instant concentration of 30 μM CO in solution 5 min prior to the scan shown in panel F. The dashed arrows indicate the direction of scanning. The asterisks (*) indicate the potential above which anaerobic inactivation takes place for CrHydA1 and DdHydAB.

rate of film loss. At each H_2 concentration the electrode rotation rate was stepped between 3000 and 5000 rpm to ensure that the current (and thus the rate of reaction) was independent of rotation rate. Values of $K_M^{H_2}$ were calculated from the x-intercept of the Lineweaver−Burk plot shown in Figure 2B and are summarized in Table 1. Experiments to measure the apparent inhibition constant (K_I^{app}, defined in eq 1 where $K_M^{H^+}$ is the K_M

Table 1. Values of $K_M^{H_2}$ and K_I^{app} for CrHydA1 and CaHydA Determined at pH 6, 10 °C from the Experiments Shown in Figure 2[a]

enzyme	$K_M^{H_2}$/mM	K_I^{app}/mM
CrHydA1	0.19 ± 0.03	3.7 ± 0.6
CaHydA	0.46 ± 0.06	6.2 ± 1.1

[a] $K_M^{H_2}$ and K_I^{app} were calculated at −0.05 and −0.4 V, respectively.

for binding of the substrate H⁺ and K_1 is the real inhibition constant) were conducted at −0.4 V vs SHE using a similar experimental method (Figure 2C). Values of K_1^{app} were calculated by adapting the method based on that reported by Léger et al.[45] which requires plotting the data according to eq 2 (in which i_{N_2} is the current recorded under 100% N_2 and i_{H_2} is the current recorded at each concentration of H_2) as shown in Figure 2D. The K_1^{app} values are included in Table 1.

$$K_1^{app} = \frac{K_1[H^+]}{K_M^{H^+}}\left(1 + \frac{K_M^{H^+}}{[H^+]}\right) \quad (1)$$

$$(i_{N_2}/i_{H_2}) - 1 = \frac{[H_2]}{K_1^{app}} \quad (2)$$

The $K_M^{H_2}$ values are much higher than those we have measured recently for some O_2-tolerant [NiFe]-hydrogenases[46] and they suggest that the catalytic current for H_2 oxidation will be sensitive to changes in H_2 levels even close to 1 bar partial pressure (this is particularly so for CaHydA). The results showed that headspace H_2 levels should be maintained constant in quantitative H_2 oxidation experiments. In contrast, the K_1^{app} values are so high that the presence of H_2 is not expected to pose a problem in studies of H_2 production. The order of magnitude difference between K_1^{app} and $K_M^{H_2}$ suggests immediately that the electronic/catalytic state of the H-cluster exerts a strong effect on binding affinity, with H_2 binding more weakly to a more reduced state.

Reactions with CO. Cyclic voltammograms of the electrocatalytic activities of DdHydAB, CrHydA1 and CaHydA at pH 6.0 under a flow of 100% H_2 are shown in Figure 3 (A, C, E—left-hand column). These voltammograms show that all the enzymes are bidirectional, with CaHydA being particularly biased in the direction of H_2 production. The traces in either potential direction cut through the zero current axis at the cell potential (see vertical line) for the 2H⁺/H_2 couple. At high potentials, all three enzymes undergo inactivation to give a species that is most likely H_{ox}^{inact} although this process is very slow for CaHydA. The voltammograms in the right-hand column (B, D, F) were recorded during the efflux of CO that had been introduced by injecting a saturated solution (giving a concentration of 100 μM CO in the cell for the experiment on CrHydA1, 30 μM for the experiment on DdHydAB) or flowing 100% CO briefly through the cell (for CaHydA) as the potential was cycled between −0.55 and 0 V prior to the scans shown in panels B, D, and F.

Informative changes in the voltammograms of DdHydAB, CrHydA1 and CaHydA occur upon removal of CO during continuous cycling (B, D, F). In all cases the voltammograms show that once the potential is made sufficiently positive to start H_2 oxidation, there is an initial increase in current which is followed by a decrease (see the dashed arrow indicating the oxidative sweep). The magnitude of this effect is most apparent after a certain time has elapsed, dependent upon the hydrogenase, and finally the voltammograms resume the expected shapes for H_2 oxidation and production at pH 6 under 1 bar H_2, analogous to those shown in the left-hand column. This cyclic 'inhibitor-on/inhibitor-off' behavior shows that CO binding to the active site is favorable and fast during H_2 oxidation but relatively weak during H_2 production. Thus CO reinhibits strongly as the potential is raised to oxidize H_2, which addresses a more oxidized form of the enzyme (above −0.3 V). The

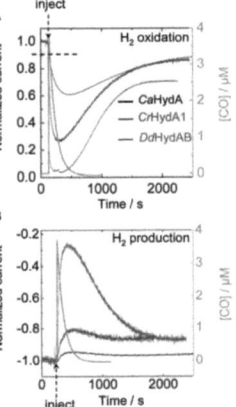

Figure 4. Inhibition of DdHydAB (red lines), CrHydA1 (blue lines) and CaHydA (black lines) by injection of 4 μM CO at (A) −0.05 V (H_2 oxidation) and (B) −0.4 V (H_2 production). Experimental conditions: pH 6.0, 10 °C, 1 bar H_2, electrode rotation 3000 rpm. The final level of the current reached upon recovery of H_2 oxidation activity of CaHydA is marked by the dashed line in A. The exponential decrease in concentration of dissolved CO is shown by the gray trace (details given in Supporting Information, Figure SI.2).

voltammogram for CrHydA1 also shows an oxidation peak soon after commencing the scan in the negative direction. A similar observation was reported in previous experiments on DdHydAB where it arises from the reactivation of some H_{ox}^{inact} and its rapid inactivation by CO.[47]

Further insight into CO binding and release under different conditions is provided by Figure 4. The CO inhibition profiles for DdHydAB, CrHydA1 and CaHydA were obtained for H_2 oxidation (Panel A) or H_2 production (Panel B) by injecting an aliquot of CO-saturated solution and then recording the catalytic current as the CO is removed by flushing. The same experimental conditions (pH 6.0, 10 °C, 1 bar H_2, electrode rotation 3000 rpm) were used for all experiments. In each case, injection of CO-saturated buffer gives an immediate initial CO concentration of 4 μM which decreases exponentially to zero,[45] as depicted by the gray trace (right y-axis) which represents the dependence of [CO] on time throughout the experiment; < 0.2 μM CO remains in solution after 500 s. (See Supporting Information (Figure SI.2) for how this dependence is determined.) In the case of CaHydA, inhibition continues to increase even though most of the CO has been flushed out of the cell, indicating that the rate of reaction with CO is slow. For H_2 oxidation, the rates and extent of inhibition reached after CO injection decrease in the order DdHydAB > CrHydA1 > CaHydA. The reactivation rates are slow and strikingly similar for all enzymes. For H_2 production (panel B) the rates and extent of inhibition by CO again decrease in the order DdHydAB > CrHydA1 > CaHydA.

(46) Ludwig, M.; Cracknell, J. A.; Vincent, K. A.; Armstrong, F. A.; Lenz, O. *J. Biol. Chem.* **2009**, *284*, 465−477.
(47) Parkin, A.; Cavazza, C.; Fontecilla-Camps, J. C.; Armstrong, F. A. *J. Am. Chem. Soc.* **2006**, *128*, 16808−16815.

RESULTS

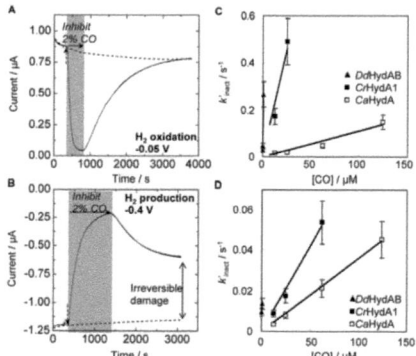

Figure 5. Inhibition by CO and reactivation, observed for H_2 oxidation and H_2 production with DdHydAB, CrHydA1 and CaHydA. Panels A and B show experiments performed on CaHydA at -0.05 and -0.4 V, respectively, in which the headgas is 80% H_2, 20% N_2 at the start of the experiment. Carbon monoxide (2%, approximately 25 μM) was introduced by injection of CO-saturated buffer (0.5 mL of buffer saturated with 10% CO, 80% H_2, 10% N_2) into the 2 mL of buffer already present in the cell and, simultaneously, flushing the headspace with 80%H_2, 18% N_2, 2% CO for the period of time marked by the gray boxes. Experimental conditions: pH 6.0, electrode rotation 3000 rpm, potentials as indicated. Panels C and D show the dependencies of rates (k'_{inact}) on CO concentration; data are derived from experiments such as those shown in panels A and B, respectively, for DdHydAB, CrHydA1 and CaHydA.

To obtain the kinetics of CO binding for the three [FeFe]-hydrogenases it was necessary to maintain a constant CO concentration throughout the time scale of the reaction. This was achieved by injecting an aliquot of solution containing CO and simultaneously changing the gas composition reaching the headspace by replacing, with CO, a certain fraction of the 20% N_2 component of the 80% H_2 mixture. Exemplary results obtained with CaHydA are shown in Figure 5. Panel A shows the time-course for CO inhibition of H_2 oxidation at -0.05 V, and Panel B shows the same time-course for inhibition of H^+ reduction at -0.4 V. Both experiments commenced with the hydrogenase on the electrode being exposed to an atmosphere of 80% H_2, 20% N_2. At $t = 300$ s, an aliquot of CO-containing solution was injected to take the cell concentration to 2% CO, and the cell was flushed with 80% H_2, 18% N_2, 2% CO. Once the catalytic current had decreased to a steady level, the CO was removed from the cell by flushing with 80% H_2, 20% N_2. (Note that 80% H_2 remained in the headspace of the cell throughout the experiment. This excludes variations in catalytic current resulting simply from changes on H_2 concentration, which would otherwise pose a problem, given the high K_M of these enzymes as indicated in Panel 1.) Pseudo first-order rate constants (k'_{inact}) were determined for each experiment from the slope of plots of ln(i) vs t, with linearity generally exceeding 75% for H_2 oxidation, although less so for H_2 production. Panels C and D show the corresponding dependencies of k'_{inact} on CO concentration, obtained for DdHydAB, CrHydA1 and CaHydA by numerous experiments conducted analogously to those shown in panels A and B.

Values for the equilibrium inhibition constant K_1^{CO}(equil) were determined for CrHydA1 and CaHydA by measuring the H_2 oxidation current stabilized at different concentrations of CO. The term K_1^{CO}(equil) depends on H_2 concentration and potential, but we kept these variables constant (80% H_2, -0.05 V). These titrations were similar in design to those shown in Figure 2C and are described in more detail in Supporting Information (Figure SI.3). Values of K_1^{CO}(equil) are shown in Table 2. Comparable data for DdHydAB could not be obtained because the enzyme was fully inhibited at the lowest practical CO concentrations.

Panel B of Figure 5 shows that CO inhibition of H_2 production of these enzymes is only partially reversible. This observation was consistently made with CaHydA and CrHydA1 but the instability of DdHydAB films prevented us from making similar measurements with this enzyme. By comparison, CO inhibition is fully reversible when measuring H_2 oxidation at -0.05 V. In all cases, the background decrease in current that we refer to as film loss (traced by the dashed line) was checked through control experiments carried out without introducing CO. The rates determined for the inactivation of H_2 production by CO are therefore approximate. The potential dependence of the rate of inactivation of H_2 production was not investigated further, although we noted that the inactivation process became noticeably biphasic as the potential was lowered below -0.4 V. We raise this issue later in the Discussion.

Panels C and D of Figure 5 show that: (i) DdHydAB is always the fastest to react with CO and CaHydA the slowest, and ii) for all three enzymes, inhibition of H_2 oxidation measured at -0.05 V is substantially faster than inhibition of H_2 production at -0.4 V, but the same order DdHydAB > CrHydA1 > CaHydA is observed in both catalytic directions. The second-order rate constants (k_{inact}) are provided in Table 2 (see later). In all cases, the reactivation rates ($k_{\text{re-act}}$) are the same, within reasonable error, for H_2 oxidation and H_2 production. Experiments carried out with CrHydA1 and CaHydA showed that $k_{\text{re-act}}$ is strongly light sensitive (see Supporting Information, Figure SI.1) in agreement with earlier electrochemical observations made with DdHydAB.[47] In addition, $k_{\text{re-act}}$ increased significantly when the temperature was raised to 25 °C. From the ratio of rate constants for the reactivation and inactivation reactions at 10 °C, we derived the kinetic inhibition constants K_1^{CO}(kin) = $k_{\text{re-act}}/k_{\text{inact}}$. In Table 2, K_1^{CO}(kin) values are compared with the equilibrium values, K_1^{CO}(equil), obtained by titration for CrHydA1 and CaHydA. The two values for CrHydA1 are in good agreement, although for CaHydA, K_1^{CO}(equil) is rather higher than K_1^{CO}(kin). The ratio of the k_{inact} values for CO inhibition of H_2 oxidation and H_2 production—the "catalytic direction discrimination"- is about 49 for DdHydAB, 22 for CrHydA1 and 3 for CaHydA: these variations clearly arise from differences in the rate that CO binds because $k_{\text{re-act}}$ is similar for all three enzymes.

Reactions with O_2. Figure 6 shows experiments in which each enzyme is subjected to 5% O_2 during H_2 oxidation at -0.05 V vs SHE. As before, the H_2 concentration was kept constant throughout the entire time-course. Each experiment began with 80% H_2, 20% N_2 flushing through the cell headspace. The gas mixture was then switched to 80% H_2, 15% N_2, 5% O_2 for the duration of the reaction (this involved simultaneous injection of an aliquot of gas-equilibrated buffer for DdHydAB and CrHydA1; the kinetics of O_2 inactivation of CaHydA were so slow that the injection was unnecessary). The gas was finally changed back to 80% H_2, 20% N_2 once most of the activity (>90%) had been eliminated. We noted that the rate of inactivation by O_2 showed a dependence on H_2 concentration, with inactivation occurring more rapidly at lower levels, thus

RESULTS

Table 2. Compilation of k_{inact} and $k_{\text{re-act}}$ Values for CO Inhibition of H_2 Production and H_2 Oxidation for DdHydAB, CrHydA1 and CaHydA, with the Corresponding Catalytic Direction Discrimination Factors and Values of K_i^{CO} Determined by Kinetic and Equilibrium Methods[a]

	DdHydAB	CrHydA1	CaHydA
H_2 oxidation $k_{\text{inact}}/s^{-1}\mu M^{-1}$	3.9×10^{-1} $\pm 3 \times 10^{-1}$	1.9×10^{-2} $\pm 1 \times 10^{-2}$	1.1×10^{-3} $\pm 7 \times 10^{-4}$
H_2 production $k_{\text{inact}}/s^{-1}\mu M^{-1}$	8.0×10^{-3} $\pm 7 \times 10^{-3}$	8.4×10^{-4} $\pm 2 \times 10^{-4}$	3.6×10^{-4} $\pm 1 \times 10^{-4}$
Catalytic direction discrimination	49	22	3
H_2 oxidation $k_{\text{re-act}}/s^{-1}$	$1.9 \times 10^{-3} \pm 2 \times 10^{-4}$	2.3×10^{-3} $\pm 1 \times 10^{-3}$	1.8×10^{-3} $\pm 1 \times 10^{-3}$
H_2 production $k_{\text{re-act}}/s^{-1}$	2.7×10^{-3} $\pm 1 \times 10^{-3}$	1.2×10^{-3} $\pm 3 \times 10^{-3}$	2×10^{-3} $\pm 2 \times 10^{-4}$
K_i^{CO} (equil)/μM for H_2 oxidation (at -0.05 V)	—	1.0×10^{-1}	2.2×10^{-1}
K_i^{CO} (kin)/μM for H_2 oxidation (at -0.05 V)	4.8×10^{-3}	1.2×10^{-1}	1.6
K_i^{CO} (kin)/μM for H_2 production (at -0.40 V)	0.34	1.4	5.6

[a] All values pertaining to H_2 production were measured at -0.40 V.

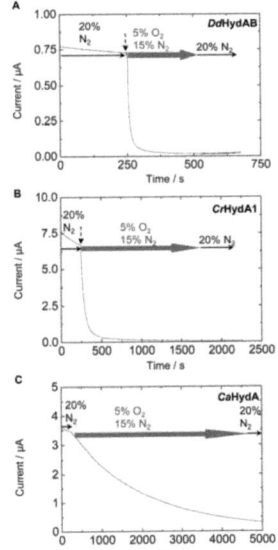

Figure 6. Inactivation of H_2 oxidation by DdHydAB, CrHydA1 and CaHydA by 5% O_2. The headspace mixture is composed of 80% H_2 and the remaining 20% is as indicated. For DdHydAB and CrHydA1 a 0.67 mL aliquot of 20% O_2/80% H_2-saturated buffer was injected into the cell which initially contained 2 mL of buffer, to give an instant concentration of 5% O_2 at the points marked by the dashed arrows. For CaHydA, the reaction was sufficiently slow that it could be initiated simply by changing the headgas mixture. Other conditions: pH 6.0, 10 °C, electrode rotation 3000 rpm, -0.05 V.

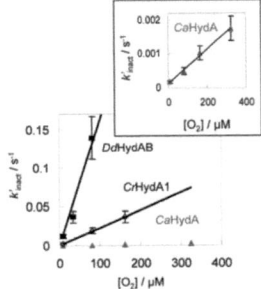

Figure 7. Dependence of rate constants (k'_{inact}) for inactivation of enzymatic H_2 oxidation by O_2, on O_2 concentration, for DdHydAB, CrHydA1 and CaHydA (shown as an expanded scale). Experimental conditions: pH 6.0, 10 °C, electrode rotation 3000 rpm, -0.05 V vs SHE, headspace gas mixture composed of 80% H_2, 20% mixture of N_2 and O_2.

indicating that O_2 and H_2 are competitive.[34] We also noted that after removing O_2 from the cell, a very small amount of activity was consistently recovered for all three enzymes. Oxygen undergoes very slow reduction at graphite at -0.05 V, therefore control experiments (Figure SI.4, Supporting Information) were performed to assess the contribution to the current from O_2 reduction. This contribution to the current was then subtracted from the experiments to verify the end point and the activity remaining. In separate anaerobic experiments we determined that small quantities of hydrogen peroxide that would be formed during electrodic O_2 reduction did not cause inactivation, by injecting aliquots of H_2O_2 solution and monitoring the effect on the H_2 oxidation current at -0.05 V (10 °C). This decrease in current was observed to be much slower than that due to equivalent concentrations of O_2 for all three enzymes.

Figure 7 shows how rate constants for O_2-inactivation of H_2 oxidation activity vary with O_2 concentration for all three [FeFe]-hydrogenases. In each case the rate of inactivation is first-order in O_2 concentration. Table 3 shows the rate constants for inactivation by O_2 alongside the rate constants for inhibition of H_2 oxidation by CO. For each enzyme, the rate of reaction with CO is much faster (80–200-fold) than with O_2 and this ratio, which we refer to as the "gas identity discrimination", is also included in Table 3.

RESULTS

Table 3. Comparison of the Second-Order Rate Constants for Inhibition by CO ($k_{inact}(CO)$) and Inactivation by O_2 ($k_{inact}(O_2)$) of DdHydAB, CrHydA1 and CaHydA and Evaluation of the Gas Identity Discrimination—the Ratio $k_{inact}(CO)/k_{inact}(O_2)$

enzyme	$k_{inact}(CO)/s^{-1}\mu M^{-1}$	$k_{inact}(O_2)/s^{-1}\mu M^{-1}$	$k_{inact}(CO)/k_{inact}(O_2)$
DdHydAB	3.9×10^{-1}	$1.8 \times 10^{-3} \pm 3 \times 10^{-4}$	217
CrHydA1	1.9×10^{-2}	$2.2 \times 10^{-4} \pm 1 \times 10^{-4}$	86
CaHydA	1.1×10^{-3}	$5.1 \times 10^{-6} \pm 1 \times 10^{-6}$	216

H_2 Production in the Presence of O_2. Finally, we carried out experiments to estimate the extent to which the rate of inactivation by O_2 depends on whether the hydrogenase is operating in the direction of H_2 production or H_2 oxidation. We used a procedure described recently,[44] in which the problem of distinguishing the current due to enzymatic H^+ reduction from that due to electrochemical O_2 reduction is resolved by adding an inhibitor. The decrease in current observed when the inhibitor is added provides a direct measure of the component of the current due to enzyme-catalyzed H^+ reduction. The experiments conducted for DdHydAB, CrHydA1 and CaHydA are shown in Figure 8.

The experiment performed on DdHydAB started under an atmosphere of N_2, and a current corresponding to enzyme-catalyzed H_2 production was recorded (stage 1). The headgas was then switched to 1% CO in N_2, and a rapid and almost complete loss of current was observed. When the CO was flushed from the cell, the current increased and reached a steady level (albeit not the same level as prior to CO introduction, due to the partial irreversibility of this reaction, see Figure 4). The current that was recovered during CO efflux was adopted as the normalization unit for the next stage. In stage 2, O_2 was introduced: the current initially increased due to direct reduction of O_2 at the graphite electrode but then began to decrease as the enzyme became inactivated. Introduction of 1% CO after 2000 s under O_2 resulted in a rapid loss of current, the magnitude of which reports on the enzyme-catalyzed current component prior to CO inhibition. Removal of CO caused the current to increase again, but a further decrease in current was observed, as expected, when O_2 was removed from the cell after $t = 8500$ s. Finally in stage 3 the inhibition step with CO was repeated to establish the extent of survival of the hydrogenase. Similar sequences of steps were used in the experiments with CrHydA1 and CaHydA, except that 99% and 95% CO, respectively, were used instead of 1% CO in order to compensate for the much slower kinetics and lower CO affinity of these hydrogenases compared to DdHydAB; in addition, 5% O_2 was used to obtain a higher rate of inactivation for CaHydA. To estimate the half-life for inactivation in each case, the decrease in current upon addition of CO after reaction with O_2 for t seconds (during stage 2) was divided by the original increase in current observed when CO was removed during stage 1. The half-life was calculated using $t_{1/2} = -t \ln 2/\ln(x)$ where x is the fraction of H_2 production current surviving after t seconds. From control experiments such as those previously described,[44] it was estimated that the concentrations of O_2 experienced by the enzyme at −0.4 V under headgas conditions of 1% and 5% were 0.7% and 4.5%, respectively. The corresponding half-lives for H_2 oxidation activity under similar O_2 concentrations were calculated from the rate constants in Table 3. The results and comparisons are shown in Table 4.

All enzymes remained at least 30% active after 2000 s under 1% bulk O_2 (i.e., at least 0.7% O_2 at the electrode[44]) for DdHydAB and CrHydA1 and even under 5% bulk O_2

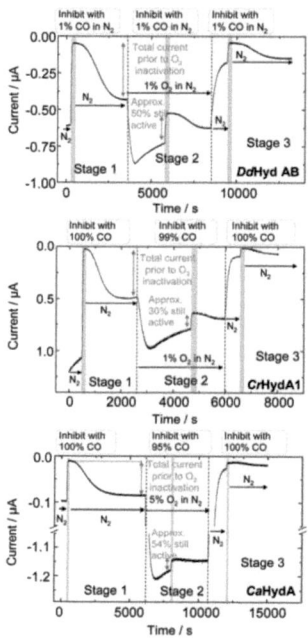

Figure 8. Chronoamperometric experiments designed to measure the survival of H_2 production activity in the presence of headspace levels of 1% O_2 for DdHydAB and CrHydA1, and 5% O_2 for CaHydA (note change in current scale upon introduction of 5% O_2). All procedures were carried out using 1 bar pressure of the gases indicated. Other conditions: pH 6.0, 10 °C. electrode rotation 3000 rpm, −0.4 V vs SHE. The blue vertical lines show introductions of CO to inhibit and reveal enzyme-catalyzed H_2 production. The fraction of enzyme surviving O_2 after time t is given by the ratio of CO-sensitive current measured in Stage 2 relative to that measured after the recovery in Stage 1 (indicated by vertical double-headed arrows in each case).

Table 4. Comparative Half-Lives (s) for O_2-Inactivation of H_2 Production and H_2 Oxidation of DdHydAB, CrHydA1 and CaHydA[a]

catalytic direction	DdHydAB	CrHydA1	CaHydA
H_2 oxidation	30	305	1890
H_2 production	~2000	~1320	~2250
Catalytic direction discrimination	60–70	4–5	approximately 1

[a] The half-lives for the H_2 production reaction are estimated on the basis of the percentage of activity remaining after an extended period (1800–2000 s, different for each enzyme) of catalytic turnover in the presence of O_2. Experimental conditions for H_2 production: DdHydAB, CrHydA1; 1% O_2 in the headgas (i.e. at least 0.7% O_2 surviving at the electrode); CaHydA, 5% O_2 in the headgas (at least 4.5% surviving at the electrode) −0.4 V, pH 6, 10 °C, electrode rotation rate 3000 rpm. Data for H_2 oxidation were calculated using the second-order rate constants (see Table 3) and values of 0.7% O_2 for DdHydAB and CrHydA1; and 4.5% O_2 for CaHydA.

(around 4.5% O_2 at the electrode) for CaHydA. More significantly, there were important differences among the three enzymes when comparing their survival to O_2 exposure during H_2 production with the data obtained for H_2 oxidation. In the case of DdHydAB, the H_2-production activity remaining is about sixty-fold greater than expected on the basis of the results described above for O_2 inactivation of H_2 oxidation. The enhancement is also observed for CrHydA1, but to a lesser extent than DdHydAB. On the other hand, no clear difference was observed for CaHydA between the rates of inactivation observed when monitoring H_2 production at -0.4 V or H_2 oxidation at -0.05 V.

Discussion

The three hydrogenases we have investigated include two with potential applications in large-scale H_2 production (CrHydA1 for photosynthesis and CaHydA for fermentation) and one of known crystal structure (DdHydAB).[48] In addition, CaHydA is closely related to CpI for which the structure is known.[22,23] Some important comparisons have been made, exploiting the unique ability of protein film electrochemistry to measure, simultaneously, the rates and extent of changes in catalytic activities under well-defined potentials (driving force). A summary of the quantitative observations and interrelationships is provided in Figure 9.

The reactions with CO are highly informative, and we[34] and others[39] have noted that CO protects [FeFe]-hydrogenases against O_2 degradation, suggesting both inhibitors target the same site. In all cases we could use CO as a strong inhibitor of both H_2 oxidation and H_2 production, helped by the fact that binding of H_2 under both conditions is much weaker than CO binding[49] (see Table 1).

Light sensitivity of CO inhibition is a well-established property of [FeFe]-hydrogenases[13,30,33,47,50-53] and originates from the photolability of the Fe-CO bond.[54] A particularly useful result is the similarity in the rates of *dark* reactivation of the CO-inhibited hydrogenases. In all cases, the rate is accelerated by illumination (as reported in an earlier study for DdHydAB[47]), and this suggests strongly that the reaction being observed in all cases is an elementary dissociation of the Fe-CO bond. This result demonstrates an intrinsic property of the H-cluster, maintained regardless of the slightly differing protein environments among the [FeFe]-hydrogenases. Evidently, all that is required for reactivation is to liberate the coordination site and ensure that CO escapes from the pocket before it can recombine.[30]

Our K_1^{CO} data for CaHydA determined during H_2 oxidation lie broadly in the range of values (around 1 μM) obtained by Thauer and co-workers[50] for CO binding to the related enzyme from *C. pasteurianum*, although those experiments also used a higher temperature and we found a consistently higher value

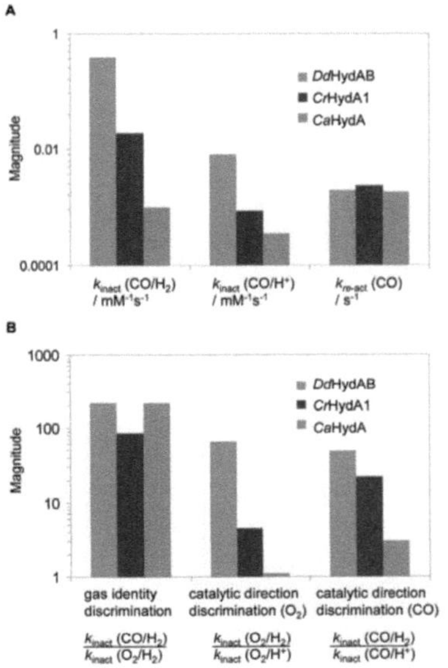

(48) Nicolet, Y.; Piras, C.; Legrand, P.; Hatchikian, C. E.; Fontecilla-Camps, J. C. *Struct. Folding Des.* **1999**, *7*, 13–23.
(49) This allowed for the same concentration of H_2 to be employed in the experiments at -0.4 and -0.05 V, thus ensuring that the experiments were comparable. For other hydrogenases, strong H_2 inhibition of H_2 production at -0.4 V would have prevented experiments performed at this potential from being performed under 80% H_2 as they were at -0.05 V.
(50) Thauer, R. K.; Kaufer, B.; Zahringe, M.; Jungerma, K. *Eur. J. Biochem.* **1974**, *42*, 447–452.
(51) Purec, L.; Krasna, A. I.; Rittenberg, D. *Biochemistry* **1962**, *1*, 270–275.
(52) Kempner, W.; Kubowitz, F. *Biochem. Z.* **1933**, *257*, 245.
(53) Albracht, S. P. J.; Roseboom, W.; Hatchikian, E. C. *J. Biol. Inorg. Chem.* **2006**, *11*, 88–101.
(54) Kochanski, E. *Photoprocesses in Transition Metal Complexes, Biosystems and Other Molecules: Experimental and Theory*; Kluwer Academic Press: London, 1992.

Figure 9. Bar charts representing various comparisons between DdHydAB, CrHydA1, and CaHydA. (A) Comparative rates of CO-inhibition of H_2 oxidation ($k_{inact}(CO/H_2)$), H^+ reduction ($k_{inact}(CO/H^+)$) and rates of recovery from CO-inhibition ($k_{re-act}(CO)$). (B) Discrimination factors characterizing the favorability of binding CO over O_2 ($k_{inact}(CO/H_2)/k_{inact}(O_2/H_2)$), binding O_2 when the enzyme is catalyzing H_2 oxidation compared to H_2 production ($k_{inact}(O_2/H_2)/k_{inact}(O_2/H^+)$) and binding CO when the enzyme is catalyzing H_2 oxidation compared to H_2 production ($k_{inact}(CO/H_2)/k_{inact}(CO/H^+)$) for DdHydAB, CrHydA1 and CaHydA. As the rate of reactivation from CO inhibition is essentially independent of the process being catalyzed, it is simply represented by the term $k_{re-act}(CO)$. The $k_{inact}(O_2/H_2)/k_{inact}(O_2/H^+)$ ratios are approximate because the values of $k_{inact}(O_2/H^+)$ are estimates. This ratio is approximated to 1 for CaHydA.

(weaker binding) for the kinetic compared to the equilibrium value. The reasons for this are unclear at present, and doubtless the model we now discuss is oversimplified. The rate data for all three enzymes, both qualitative (Figure 3) and quantitative (Figure 4 and Figure 5), reveal a preference for CO binding to the enzymes under H_2 oxidation—the direction enforced by a higher electrode potential that should ensure that H_{ox} predominates over H_{red} during the catalytic cycle. The "catalytic direction discrimination" for CO decreases in the order DdHydAB > CrHydA1 ≫ CaHydA—the same order as found for the equivalent factor estimated for O_2 inactivation. The order also matches that observed for the rates of inactivation of H_2 oxidation by CO and O_2. These interrelationships lead us to propose a model for the *reversible* binding of CO to [FeFe]-hydrogenases which can be extended to account for reactions with O_2. The model is represented schematically in Figure 10A.

The model considers that the attack by CO (rate constant k_{inact}) involves two stages: the first stage is transport of CO (we treat this generically as the small molecule X) from the external medium (X_{ext}) through the enzyme (for simplicity we show this as a single tunnel without branches or sites at which X could

RESULTS

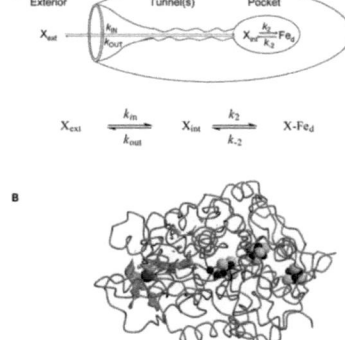

Figure 10. (A) Cartoon depicting the stepwise reaction of an inhibitory gas molecule X (= CO) with the buried H-cluster (Fe$_d$) of a [FeFe]-hydrogenase, showing transfer from the external medium (X$_{ext}$) to a position (X$_{int}$) close to Fe$_d$ (k_{in}, k_{out}), and inner-sphere coordination/dissociation at Fe$_d$ (k_2, k_{-2}). (B) C-α tracing of the [FeFe]-hydrogenase from *D. desulfuricans*: *Dd*HydAB(large subunit: blue and green, small subunit: red) and the cavity observed in the crystals (probe size: 0.8 Å, program CAVsel, A. Volbeda, unpublished). The H-cluster occupies the center of the molecule. The blue sphere represents the experimentally observed Xe site. Other color codes: red: Fe, yellow: S.

be trapped) to a noncoordinating site close to the H-cluster; the second stage is migration of X$_{int}$ from the noncoordinating site to a coordination site that we assume to be Fe$_d$ (the actual inner-sphere binding reaction). The process of reactivation (overall rate constant k_{re-act}) is the reverse of this reaction scheme.

The terms k_{in} and k_{out} are rates of transport of X through the protein in either direction, and k_2 and k_{-2} are the elementary rates of binding and dissociation of ligand X, respectively, to Fe$_d$ within the region of the active-site pocket. With the simplification that $k_{-2}[X\text{-Fe}_d]$ is negligible until the reaction of inhibitor with enzyme is essentially complete, the steady-state approximation with $d[X_{int}]/dt = 0$ yields:

$$\text{rate of inactivation} = \frac{k_{in}k_2[X_{ext}][Fe_d]}{k_{out} + k_2} \quad (3a)$$

where the pseudo first-order rate constant (as measured directly in experiments) is:

$$k'_{inact} = \frac{k_{in}k_2[X_{ext}]}{k_{out} + k_2} \quad (3b)$$

An assumption of this model is that k_{in} and k_{out} should depend on the nature of the gas molecule and gas filter but not on the catalytic state of the H-cluster—the state predominating for a particular electrode potential, i.e. H$_{ox}$ or H$_{red}$. Based on the evidence that CO binds preferentially (and perhaps exclusively) to H$_{ox}$,[29] we expect that k_2 for CO will be large for H$_{ox}$ and small, even zero, for H$_{red}$. Note however that inhibition is still observed during H$_2$ production because H$_{ox}$ may always appear briefly in the catalytic cycle, even at the lowest potentials we have used (−600 mV in the cyclic voltammetry experiments).

Recognizing this is a simplistic model, we now consider the following limiting scenarios: (i) if $k_{out} \leq k_2$, $k'_{inact} \sim k_{in}[X_{ext}]$ so the rate of inhibitor binding depends only on the external concentration and rate of internal transport of X; in this case little discrimination is expected based on the redox state of the H-cluster. Alternatively, (ii), if $k_{out} \gg k_2$, i.e. if the protein's internal structure does not provide an effective barrier to transport of X, it follows that $k'_{inact} = k_2 k_{in}[X_{ext}]/k_{out}$. In this scenario, the rate of inhibition depends not only on the nature of X but also upon k_2 and therefore should also be faster for conditions favoring H$_{ox}$ (H$_2$ oxidation) compared to H$_{red}$ (H$_2$ production).

Reactivation follows the reverse sequence, and for $k_{in}[X_{ext}] = 0$ (because CO is removed from the solution) we obtain

$$\text{rate of re-activation} = \frac{k_{out}k_{-2}[X\text{-Fe}_d]}{k_{out} + k_2} \quad (4a)$$

and the first-order rate constant (as measured experimentally) is given by

$$k_{re-act} = \frac{k_{out}k_{-2}}{k_{out} + k_2} \quad (4b)$$

Under the limiting condition $k_{out} \ll k_2$, $k_{re-act} = k_{out}k_{-2}/k_2$, whereas if $k_{out} \gg k_2$, the rate of reactivation reduces to k_{-2}, reflecting the likelihood that CO escapes from the enzyme (k_{out}) before it can recoordinate (k_2). Our data suggest that the latter situation must generally be the case, although a more intermediate situation (a smaller k_{out} relative to k_2) applying for *Ca*HydA (see below). Overall, the dissociation constant is given by $K_i^{CO}(\text{kin}) = k_{re-act}/k_{inact} = k_{out}k_{-2}/k_{in}k_2$, which always depends on the kinetics of making and breaking the Fe-CO bond.

This analysis can be extended to the reaction of [FeFe]-hydrogenases with O$_2$, although that reaction is essentially irreversible. Table 3 and Figure 9 show that trends among the hydrogenases as observed for their reactions with CO are mirrored in their reactions with O$_2$; for example, *Dd*HydAB shows the highest rates of inhibition in both cases and the greatest discrimination based on catalytic direction.

In mechanistic terms, the miniscule protection that *Dd*HydAB possesses against attack by O$_2$ is provided only within the active-site pocket in which O$_2$ is able to discriminate between different catalytic states of the enzyme (a strong k_2 dependence, according to the model). In contrast, the small catalytic direction discrimination observed for *Ca*HydA can be interpreted in terms of it showing a less excessive value of k_{out} (a more restrictive tunnel or filter) in relation to k_2. Values for k_{out} should correlate closely with those for k_{in};thus, it is significant that *Ca*HydA also shows the slowest rates of reaction with CO and O$_2$ and, with a half-life of several minutes under atmospheric O$_2$ levels at 10 °C, looks to be a promising model for aerobic biohydrogen production even though it stems from a strict anaerobe.

The evidence (strong light enhancement) that the rate-determining step in reactivation is the elementary scission of the Fe-CO bond in the H$_{ox}$-CO state, and the observation that the rate measured in the dark is quite similar for all enzymes (which share only 40% sequence similarity) shows that the kinetics of reactivation are governed more by the intrinsic properties of the H-cluster than by the nature of the surrounding enzyme. Lubitz and colleagues have proposed that the H-clusters in *Cr*HydA1 and *Dd*HydAB are similar, although they differ slightly in electronic detail.[28] Note that were CO to coordinate

RESULTS

to different states of the H-cluster at -0.4 and -0.05 V, we would expect k_{re-act} to depend significantly on potential, but it does not. This supports the view that CO (and by extension, O_2) binds to H_{ox} but not H_{red}. The other comparison in Figure 9 which is reasonably constant among all three enzymes is the gas identity discrimination (CO vs O_2). This again may reflect intrinsic behavior of the H-cluster because CO is a superior ligand to O_2 in terms of its π-acceptor capability. Dominant intrinsic effects are not unexpected, given the unusually independent status of the $2Fe_H$ subcluster, which was described in the Introduction as an enzyme cofactor resembling an organometallic compound buried in a protein. Clearly our model is an oversimplification, albeit necessary at this stage, and to ...and this observation more fully we are undertaking ...ical calculations, including predictions of relative trans-...es through the enzyme.

...notion of a filter or a tunnel connecting the H-cluster to ...lecular surface is supported by the two available [FeFe]-...enase structures. In DdHydAB and CpI there is a 'static' ...that can be revealed using a cavity-searching program.[55] ...nel in each enzyme has a central cavity that can bind a ...m and a narrower path leading to Fe_d.[17,26] Figure 10B ...the experimentally observed tunnel and Xe site in ...AB.[56,57] The tunnel connects the molecular surface to ...ve site, and one possibility is that the Xe atom occupies ... in which dissociated CO could reside before rebinding ...or escaping to the medium. Molecular dynamics simula-...ased upon the CpI structure revealed a second tunnel ...o connects to the central cavity.[58] What can be concluded ...th crystallographic and theoretical studies is that dynamic ...ions are important for intramolecular gas diffusion in ...hydrogenases.

...further mechanistic points emerge from this study. First, ...ays observed that CO inhibition of H_2 production is only ...y reversible. At present we have no explanation for this, ...h Adams reported in 1987 that CO binds irreversibly to ...tic intermediate of CpI.[59] Further investigations including ...tudy of the potential dependence of CO binding during ...duction are clearly required to resolve this issue, which ...ve mechanistic relevance. Second, we always recorded ...small proportion of activity returning after O_2 inactivation, an observation in line with those reported by Baffert et al.[49]

also in studies on CaHydA. Such a part-reversal is consistent with the mechanism proposed by Stripp et al.[34] in which O_2 must first bind in a reversible manner at the distal Fe of the $2Fe_H$ subcluster before causing irreversible damage to the $[4Fe-4S]_H$ subcluster.

From a biological perspective, the precise, quantitative data that we have been able to extract and compare for the three different hydrogenases should be understandable in terms of the lifestyles of the organisms that express them. This is true, in part. Recent studies on the O_2 detoxification mechanism in *C. acetobutylicum* have shown that this fermentative bacterium can survive limited exposure to air and can even undergo cell division at surprisingly high concentrations of O_2.[60,61] The relative O_2 stability of CaHydA may therefore be a consequence of concerted evolutionary adaptation to an O_2-rich atmosphere. However, certain species of the *Desulfovibrio* genus have also been reported to exhibit short-term survival when exposed to O_2;[62,63] thus, from a microbiological viewpoint the large disparity in O_2 sensitivity between CaHydA and DdHydAB is puzzling. The 'intermediate' degree of O_2 sensitivity displayed by CrHydA1 is consistent with the observation that although it is only expressed in *C. reinhardtii* under anaerobiosis[64,65] it is likely to be in contact with at least trace amounts of O_2 that are produced by photosystem II.[66]

Acknowledgment. Research in the group of FAA was supported by the UK BBSRC (Grant BB/D52222X) and EPSRC. C.C. and J.C.F.C. thank the CEA and the CNRS (France) for institutional funding. S.S. and T.H. were supported by the Deutsche Forschungs-gemeinschaft (SFB 480) and EU/Energy Network SolarH2 (FP7 418 contract 212508. We thank Dr. Alison Parkin for providing part of Figure 3 and for helpful advice, and Dr. A. Volbeda for preparing Figure 10B.

Supporting Information Available: Photolability of the CO-bound state; determination of the variation of the concentration of CO with time in the experiments shown in Figure 4; determination of K_I^{CO}(equil); control experiments for experiments investigating O_2 inactivation of H_2 oxidation. This material is available free of charge via the Internet at http://pubs.acs.org.

JA905388J

---

(55) Montet, Y.; Amara, P.; Volbeda, A.; Vernede, X.; Hatchikian, E. C.; Field, M. J.; Frey, M.; Fontecilla-Camps, J. C. *Nat. Struct. Biol.* **1997**, *4*, 523–526.
(56) Fontecilla-Camps, J. C.; Amara, P.; Cavazza, C.; Nicolet, Y.; Volbeda, A. *Nature* **2009**, *460*, 814–822.
(57) Nicolet, Y.; Piras, C.; Legrand, P.; Hatchikian, C. E.; Fontecilla-Camps, J. C. *Structure* **1999**, *7*, 13–23.
(58) Cohen, J.; Kim, K.; Posewitz, M.; Ghirardi, M. L.; Schulten, K.; Seibert, M.; King, P. *Biochem. Soc. Trans.* **2005**, *33*, 80–82.
(59) Adams, M. W. W. *J. Biol. Chem.* **1987**, *262*, 15054–15061.

(60) Hillmann, F.; Fischer, R. J.; Saint-Prix, F.; Girbal, L.; Bahl, H. *Mol. Microbiol.* **2008**, *68*, 848–860.
(61) May, A.; Hillmann, F.; Riebe, O.; Fischer, R. M.; Bahl, H. *FEMS Microbiol. Lett.* **2004**, *238*, 249–254.
(62) Cypionka, H. *Annu. Rev. Microbiol.* **2000**, *54*, 827–848.
(63) Lumppio, H. L.; Shenvi, N. V.; Summers, A. O.; Voordouw, G.; Kurtz, D. M. *J. Bacteriol.* **2001**, *183*, 2970–2970.
(64) Happe, T.; Naber, J. D. *Eur. J. Biochem.* **1993**, *214*, 475–481.
(65) Happe, T.; Kaminski, A. *Eur. J. Biochem.* **2002**, *269*, 1022–1032.
(66) Happe, T.; Hemschemeier, A.; Winkler, M.; Kaminski, A. *Trends Plant Sci.* **2002**, *7*, 246–250.

RESULTS

2.6 How Algae produce Hydrogen – News from the photosynthetic Hydrogenase

Sven T. Stripp & Thomas Happe*

Lehrstuhl Biochemie der Pflanzen, AG Photobiotechnologie, Ruhr Universität Bochum, Universitätsstrasse 150, 44801 Bochum, Germany

*Corresponding Author:
Phone: +49-(0)234-32 27026; Fax: +49-(0)234-32 14322
E-mail: thomas.happe@rub.de

http://dx.doi.org/ 10.1039/b916246a

PERSPECTIVE

How algae produce hydrogen—news from the photosynthetic hydrogenase

Sven T. Stripp and Thomas Happe*

Received 7th August 2009, Accepted 9th September 2009
First published as an Advance Article on the web 22nd October 2009
DOI: 10.1039/b916246a

Green algae are the only known eukaryotes capable of oxygenic photosynthesis which are equipped with a hydrogen metabolism. Hydrogen production is light-dependent, since the [FeFe] hydrogenases are coupled to the photosynthetic electron transport chain *via* ferredoxin. Algal [FeFe] hydrogenases are one of the most active biocatalysts for the evolution of hydrogen. Therefore, special interest exists in the biophysical characterization and biotechnological usage of these [Fe-S] enzymes. This review traces the discovery of this interesting class of proteins. Recent findings allow insight into the electronic structure and configuration of the [FeFe] hydrogenase active site (H-cluster). Emphasis is placed on novel discoveries of the hydrogenase interaction with its natural electron donor ferredoxin and the mechanism of enzyme inactivation through oxygen.

Introduction

Hydrogenases catalyze a simple reaction, namely the reversible reduction of protons to molecular hydrogen. The discovery of this class of enzymes was made in the 1930s.[1] Years later, Hans Gaffron observed that green algae can either oxidize hydrogen in concert with CO_2 fixation in the "dark reaction"[2,3] or evolve hydrogen gas upon illumination.[4] Since this important finding, the hydrogenase metabolism in photosynthetic algae has been of great scientific interest. Stuart and Gaffron were the first to uncover the direct links between hydrogen evolution and photosynthesis,[5] and in the late 1990s, Melis and co-workers established sulfur deprivation for semi-continuous, photobiological hydrogen production in *Chlamydomonas reinhardtii*.[6]

This breakthrough towards a sustainable hydrogen production was achieved by separating oxygenic photosynthesis and CO_2

Lehrstuhl Biochemie der Pflanzen, AG Photobiotechnologie, Ruhr Universität Bochum, Universitätsstrasse 150, 44801, Bochum, Germany. E-mail: thomas.happe@rub.de; Fax: +49-(0)234-32 14322; Tel: +49-(0)234-32 27026

fixation from hydrogen evolution in time. Wykoff and Melis could show that a sulfur-deprived culture of *C. reinhardtii* gradually loses its photosynthetic capacity while mitochondrial respiration is left essentially unchanged.[6,7] Photosynthesis is diminished due to the loss of the catalytic active D1 subunit of photosystem II (PSII) which turns over very rapidly.[7] Deprived of sulfur, the amino acids cysteine and methionine run short and D1 can not be replaced at an appropriate rate. Thus, PSII-catalyzed water oxidation and oxygen evolution decline. Once respiration consumes more oxygen than residual photosynthesis can deliver, cells become anaerobic and hydrogen turnover is induced.[8] Under sulfur deprivation, reduction of protons is a sink for (excess) electrons that result from starch breakdown as a product of CO_2 fixation during cell growth under oxygenic conditions.[9,10]

The hydrogenase HydA1 of *C. reinhardtii* receives electrons at the reducing end of the photosynthetic electron transfer chain. The "photosynthetic" ferredoxin PetF shuttles electrons from photosystem I (PSI) to HydA1 which reduces protons to molecular hydrogen.[11] The hydrogenase competes with different electron sinks, in particular ferredoxin-NADP-oxidoreductase as

Sven T. Stripp

Sven T. Stripp studied chemistry and biology at the Ruhr-University Bochum. His diploma in Biology was awarded in 2006. In his diploma thesis within the group of Prof. Thomas Happe he helped establishing a system for the heterologous synthesis of [FeFe] hydrogenases. During his PhD thesis Sven T. Stripp focuses his research on the reaction mechanism of [FeFe] hydrogenases and the oxygen sensitivity of this class of [FeS] proteins.

Thomas Happe

Prof. Thomas Happe received his PhD in 1994 from Ruhr-University Bochum. He worked as a postdoc at the University of Bonn in the Department of Molecular Biochemistry and became associate professor in 1998. In the year 2000 he was guest fellow at the University of California in Berkeley. He has been a full professor at the Ruhr-University Bochum since 2003. His research interest focuses on the photobiological hydrogen production in green algae and the characterization of [FeFe] hydrogenases.

RESULTS

an interface with the Calvin cycle.[12–14] Unlike PSII, the PSI complex is essential to hydrogen evolution.[15] "Catabolic" electrons are fed into the photosynthetic electron transfer chain from degradation of starch, glucose or acetate at the level of the plastoquinone pool.[16–18] This PSII-independent hydrogen evolution, which utilizes fermentative oxidation of organic substrates, is referred to as "photofermentation".[14,19]

Hydrogenases are ubiquitous in strict and facultative anaerobes, and the vast majority is found in prokaryotes.[20,21] Hydrogenases are transition metal enzymes, likely to be developed in a pre-photosynthetic, reducing atmosphere.[22,23] In the absence of oxygen, hydrogenases serve as terminal electron acceptors. However, hydrogenases are found in oxidation ("uptake") of molecular hydrogen as well.[24] According to the composition of the bimetallic active site cofactor, [NiFe], [FeFe] and hydrogen-forming methylenetetrahydromethanopterin (Hmd) [Fe] hydrogenases are distinguished.[25–29] Under physiological conditions, [NiFe] hydrogenases generally act as uptake hydrogenases while [FeFe] hydrogenases often catalyze hydrogen evolution.[30,31] Hydrogen release with [FeFe] hydrogenases is fast and in most cases controlled by diffusion of substrates and products.[23,25] [NiFe] hydrogenases were shown to exhibit much higher affinities for hydrogen (as a substrate in uptake) than [FeFe] hydrogenases.[24] [NiFe], [FeFe] and [Fe] hydrogenases (Hmd) are *not* homologs and give a model for convergent mechanistic evolution.[32]

[FeFe] hydrogenases have been described for pro- and eukaryotes, [NiFe] hydrogenases in contrast are solely found in prokaryotes including cyanobacteria. Green algae and cyanobacteria are the only organisms currently known to be capable of both oxygenic photosynthesis and hydrogen production.[33] However, despite the availability of a number of entirely sequenced cyanobacterial genomes, [FeFe] hydrogenases have never been described in cyanobacteria. The photosynthetic cyanophyta ("blue-green algae") are endosymbiotic progenitors of plastids that form chloroplasts in higher plants and algae.[21] These eubacteria possess only [NiFe] hydrogenases and evolve hydrogen by a light-dependent reaction. The cyanobacterial hydrogen metabolism is different to the algal hydrogenase turnover and catalyzed by nitrogenase, the nitrogen fixing enzyme complex.[33] Thus it appears that, in green algae, the hydrogenase has been introduced by a host with a nucleus-encoded [FeFe] hydrogenase of non-cyanobacterial origin.

Besides the natural [FeFe] and [NiFe] catalysts for hydrogen production, chemists have developed several electro- and photochemical hydrogen evolving catalyst systems in the last few years. Based on the [2Fe-2S] cofactor of the H-cluster from [FeFe] hydrogenases, it was shown that these structural and functional [2Fe-2S] mimics can efficiently produce hydrogen.[34–38] Moreover, hydrogen catalysts can also be coupled to photosensitizers and release hydrogen by light-induced water splitting. The progress and the application of the metal-based devices for light driven hydrogen evolution in homogeneous systems was recently summarized by Wang and co-workers.[39]

During the past fifteen years, traditional physiological and biochemical studies have yielded information on photobiological hydrogen evolution in green algae.[6,8] Several review articles summarize the discovery of hydrogen turnover under sulfur deprivation and the isolation of genes encoding for different algal hydrogenases.[11,17,18,31,40,41] The purpose of this article is to highlight the wealth of new results regarding biophysical properties of the [FeFe] hydrogenase HydA1 of *C. reinhardtii*, including the electronic structure of the active site H-cluster, the reaction of this prosthetic group with CO and oxygen, and the interaction of the algal protein with its native donor ferredoxin.

The discovery of hydrogenases in green algae

As already mentioned, the first descriptions of hydrogen evolution by photosynthetic algae were published seventy years ago by Hans Gaffron and co-workers.[4] In 1973, Eric Kessler summarized the relevant information on hydrogen production by photosynthetic algae in a review article, showing that many species of unicellular green algae are equipped for hydrogen metabolism.[42] However, thirty years elapsed between the first observation of a "Cell-free Hydrogenase from *Chlamydomonas*" by Frederick B. Abeles[43] and the purification of the *C. reinhardtii* hydrogenase by Happe and Naber in 1993.[44]

Abeles could show in his pioneering experiments that the cell-free preparations of *Chlamydomonas eugametos* evolve hydrogen when the hydrogenase fraction was incubated with reduced methylviologen as electron mediator.[43] He also analyzed the inactivation of the protein by small amounts of oxygen and carried out his experiments under strict anaerobicity. However, his observation that the hydrogenase is not associated with the chloroplast was incorrect.

Twenty years later, Paul G. Roessler and Stephen Lien developed a method which resulted in a 2000-fold purification of hydrogenase HydA1 of *C. reinhardtii*.[45] The trick was to use an affinity chromatography with immobilised ferredoxin which is the electron donor to the hydrogenase *in vivo*. The preparation was 40% pure and the specific hydrogen evolution capacity of the enzyme was calculated to be 1800 μmol H_2 min^{-1} mg^{-1}.[45] Additional experiments on HydA1 showed that "activation and *de novo* synthesis" of the protein was inhibited by cycloheximide but not chloramphenicol.[46] These results clearly indicated that the hydrogenase gene is nucleus-encoded. Roessler and Lien gave the hydrogenase research in green algae an important impulse leading to the eventual isolation of the hydrogenase from *C. reinhardtii* in the beginning of the 1990s.

To characterize the algal hydrogenase in more detail, the next step was to isolate the protein up to homogeneity. Thomas Happe and Dirk Naber used five column-chromatography steps to purify the enzyme 6100-fold and determined the specific activity for hydrogen evolution as 935 μmol H_2 min^{-1} mg^{-1}.[44] A single band was observed on SDS PAGE gels which had an apparent molecular mass of 48 kDa. The respective protein fraction on non-denaturing gels possessed methylviologen reducing activity. Another study showed that the protein contains iron but no nickel.[47] Therefore, and because of the specific biochemical properties of the enzyme (CO inhibition, extreme oxygen sensitivity; see below), the authors classified the algal hydrogenase as [FeFe] hydrogenase (originally "Hydrogenase of the Fe-only type").

During that time in the 1990s, the results of Happe and Naber were called into question because it was known that cyanobacteria, the free-living precursors of plastids, encode exclusively for [NiFe] hydrogenases. Schnackenberg *et al.* published the isolation of an ostensible [NiFe] hydrogenase of the green alga *Scenedesmus obliquus*.[48] While it was not yet established that the *C. reinhardtii*

RESULTS

hydrogenase is encoded in the nucleus,[46] the scientific community knew that the algal chloroplast phylogenetically results from endosymbiosis of cyanobacteria. Hence, doubt was sown that algae can contain any other than [NiFe] hydrogenases.

The conflict was solved when the Happe group isolated the hydrogenase gene from *C. reinhardtii*.[49] The deduced amino acid sequence of HydA1 revealed a conserved C-terminal sequence typical for [FeFe] hydrogenases, including four conserved cysteine residues that coordinate the active site.[41,50] Based on these elementary results, the *hydA* genes of further algal species were isolated in the following years.[51,52] It turned out that the hydrogenase proteins of algae represent a novel class of [FeFe] hydrogenases.[17] The "chlorophyta-type" [FeFe] hydrogenases are smaller (44–48 kDa) because they lack the N-terminal ferredoxin-like domain ("F-domain") present in all [FeFe] hydrogenases isolated back then (see below).[53] Moreover, the reported occurrence of [NiFe] hydrogenases in green algae[46] has never been supported by gene cloning and sequencing and was proven to be erroneous.

Although the genes and the proteins of algal hydrogenases were isolated, another problem had to be overcome to learn more about this class of enzymes. The problem was explained by Roessler and Lien as follows: "More detailed analysis of the active site of *C. reinhardtii* hydrogenase by the use of electron spin resonance (EPR) spectroscopy would be highly desirable for comparative purposes, but the low quantity of hydrogenase present in this organism makes this a difficult task".[45] In the early 1990s, Happe and Naber also reported that they could isolate only 1 μg protein per liter of green algae culture.[44]

To overcome these difficulties, two strategies were used. First, newly established and efficient induction and purification protocols, *e.g.* isolation of the hydrogenase from a sulfur-deprived algal culture, yielded 40 μg HydA1 from one liter green algae corresponding to a 40-fold increase in protein content compared to previous protocols.[54] Second, a heterologous expression system for [FeFe] hydrogenases in the fermentative bacterium *Clostridium acetobutylicum* was established.[55] Using *Escherichia coli* or *Shewanella oneidensis* as hosts, synthesis led to only low amounts of recombinant [FeFe] hydrogenases[56] or high amounts of protein but comparably low specific activities.[57] The heterologous synthesis of [FeFe] hydrogenases with *C. acetobutylicum* in contrast offers the possibility to produce both large amounts of enzyme *and* hydrogenase at high specific activity. After optimizing various parameters, it is possible to isolate about 2 mg of pure and active [FeFe] hydrogenase from one litre of bacterial cell culture.[58] Furthermore, side directed mutagenesis on the plasmid-encoded proteins allows the investigation of structure-function relationships in [FeFe] hydrogenases by analyzing the characteristics of [FeFe] hydrogenase variants.

Basic properties of the [FeFe] hydrogenases from green algae

[FeFe] hydrogenases are small, mono- and dimeric enzymes of 45–65 kDa. The active site cofactor is a unique [Fe-S] compound commonly referred to as "H-cluster".[59] *In vivo*, [FeFe] hydrogenases are usually found in hydrogen evolution.[30] Essayed *in situ*, catalysis is mostly bidirectional. Enzyme activity is easily inactivated by oxygen and $CO^{22,60}$ although the characteristics of inactivation differ in the reduction (evolution) and oxidation (uptake) directions.[61] [FeFe] hydrogenases from organisms like *Clostridium pasteurianum* and *acetobutylicum*, *Desulfovibrio desulfuricans* and *Megasphaera elsdenii* have been described in detail.[25,55,58,62] Table 1 shows a comparison of the basic properties of bacterial and algal hydrogenases, in particular from *C. reinhardtii*, *S. obliquus*, *Chlamydomonas moewusii*, *Chlorococcum submarinum* and *Chlorella fusca*. All [FeFe] hydrogenases are efficient catalysts in hydrogen evolution, but bacterial enzymes like *Cp*I, *Dd*H and HydA of *M. elsdenii* release hydrogen at exceptionally high rates (5000–8000 μmol H_2 min^{-1} mg^{-1}).

Most [FeFe] hydrogenases consist of a single peptide chain. The structures of [FeFe] hydrogenases of *D. desulfuricans* (*Dd*H) and *C. pasteurianum* (*Cp*I) have been resolved by X-ray crystallography.[50,59] *Cp*I represents the typical bacterial-type [FeFe] hydrogenase. Two domains can be distinguished. The C-terminal "H-domain" carries the H-cluster, an electronically coupled [6Fe-6S] cluster described in detail later on. The accessory F-domain holds a set of ferredoxin-type [4Fe-4S] and/or [2Fe-2S] clusters.[20] These [Fe-S] compounds form an electric "wire", shuttling electrons from the protein surface to the H-cluster.[61] [FeFe] hydrogenases of green algae belong to the smallest hydrogenases known and are about 15 kDa smaller than most bacterial hydrogenase enzymes. The H-cluster is the only catalytically active iron compound in algal hydrogenases. According to sequence alignment, this is true for all [FeFe] hydrogenases found in algae up to now.[52,54,63]

Table 1 Comparison of different prokaryotic and chlorophyta-type [FeFe] hydrogenases

Organism	Name	M_r^a	V_{max}^b	Reference
Clostridium pasteurianum	CpI	63.8	5500	Adams 1990 (25)
Clostridium acetobutylicum	HydA	64.3	1750	von Abendroth *et al.* 2008 (58)
Desulfovibrio desulfuricans	DdH	46.1+14.0	8820	Hatchikian *et al.* 1992 (62)
Megasphaera elsdenii	HydA	53.6	7000	Adams 1990 (25)
Chlamydomonas reinhardtii	HydA1	47.5	935	Happe and Naber 1993 (44)
Chlamydomonas reinhardtii	HydA2	47.3	n.d.	Forestier *et al.* 2003 (66)
Scenedesmus obliquus	HydA	44.6	630	Girbal *et al.* 2005 (55)
Chlamydomonas moewusii	HydA1	45.4	1600	Kamp *et al.* 2008 (54)
Chlorococcum submarinum	HydA	45.3	640	Kamp *et al.* 2008 (54)
Chlorella fusca	HydA	45.1	1000	Winkler *et al.* 2002a (17)

a M, Molecular weight in kDa as derived from protein primary structure. *b* V_{max} Specific hydrogen evolution activity expressed as μmol H_2 min^{-1} mg^{-1} with 10 mM methylviologen as electron donor; n.d. not determined.

RESULTS

Interestingly, these proteins are exclusively found in the chloroplast stroma and are not associated with the membrane, which is different to the situation reported for [NiFe] hydrogenases.[64] However, the hydrogenase genes are encoded in the nucleus and a N-terminal transit peptide allows for import of the transcript to the chloroplast where protein biosynthesis is thought to take place.[41,65] While prokaryotic hydrogenases are usually part of the fermentative metabolism, [FeFe] hydrogenases in algae receive reducing equivalents at the end of the photosynthetic electron transfer chain *via* Ferredoxin.[40,41] Therefore, chlorophyta-type [FeFe] hydrogenases have been termed "photosynthetic hydrogenases".[18]

As reported for most bacteria, an isoenzyme HydA2 was found in *C. reinhardtii* and other green algal genomes. The gene *hydA2* possesses all conserved residues and domains identified typical for the active site of this class of [FeFe] hydrogenases.[66] Up to now, the HydA2 protein has not been isolated, and the function and catalytic activity of this isoenzyme remains unclear.

Fig. 1 displays a sequence alignment for different algal hydrogenases with the structurally well-characterized H-domain of prokaryotic enzymes *Cp*I and *Dd*H. The hydrogenase of *M. elsdenii* is plotted because of its minimal set of accessory [Fe-S] clusters (see Fig. 2). For reasons of simplicity, the accessory F-domain is not shown in the alignment. For all [FeFe] hydrogenases, conservation of the active site motif is evident. The position of the H-cluster cysteines (C) is well-preserved and around these residues, sequence similarity is comparably high. However, certain differences set apart bacterial and chlorophyta-type hydrogenases. The N-terminal F-domain is missing (sequence not shown), instead all algal hydrogenases display an "insertion" region (dashed boxes 1 and 2). This insertion (most pronounced in HydA1 of *C. reinhardtii*) is discussed to form a loop responsible for the interaction with the *in vivo* electron donor Ferredoxin.[67] The binding niche for ferredoxin is formed by bulky, basic amino acid residues (K and R in Fig. 1) which are conserved in chlorophyta-type hydrogenases exclusively.

Bacterial-type [FeFe] hydrogenases are very similar to algal [FeFe] hydrogenases in terms of the H-domain primary structure. Variety exists for the accessory F-domain, which structurally differs in prokaryotic [FeFe] hydrogenases and is missing in chlorophyta-type hydrogenases. A relay of [Fe-S] clusters is associated with this domain. However, the amount of bound clusters varies from two (HydA of *M. elsdenii*) to four (*Cp*I).[50,68] Fig. 2 compares the cartoon model crystal structures of *Cp*I (1FEH) and *Dd*H (1HFE) with homology models of HydA of *M. elsdenii* and HydA1 of *C. reinhardtii*.

From Fig. 2, the functional bisection of *Cp*I is easy to see. The upper H-domain holds the H-cluster, the accessory F-domain exhibits three [4Fe-4S] clusters and one [2Fe-2S] compound ("F-clusters"). The overall shape resembles a mushroom.[50] In HydA of *M. elsdenii*, the F-domain is decreased in size. This bacterial [FeFe] hydrogenase holds only two [4Fe-4S] clusters besides the prosthetic group of the H-domain.[68] The periplasmatic *Dd*H differs in structure as the enzyme is a heterodimer and compromises two single-chain subunits, giving the overall molecular weight of approximately 60 kDa. The small 14 kDa chain is discussed to be relevant in translocation to the periplasmatic space.[59] Unlike the F-domain, this subunit does not contain any [Fe-S] clusters or respective binding motifs. However, next to the H-cluster, two [4Fe-4S] clusters are found with the 46 kDa subunit. HydA1 of *C. reinhardtii*, as a representative of chlorophyta-type hydrogenases, lacks the F-domain.[54,69] The putative binding niche of ferredoxin is marked in Fig. 2, as well as the insertion region discussed by Winkler and co-workers.[67]

Electronic structure of the H-cluster

The H-cluster is composed of a ferredoxin-type [4Fe-4S] cluster linked to a [2Fe-2S] moiety commonly known as "[2Fe]$_H$". Each iron atom of the [2Fe]$_H$ cluster is coordinated with one cyanide group (CN$^-$) and one or two carbon monoxide groups (CO).[59] In respect to the position of the [4Fe-4S] subcluster, the [2Fe]$_H$ iron

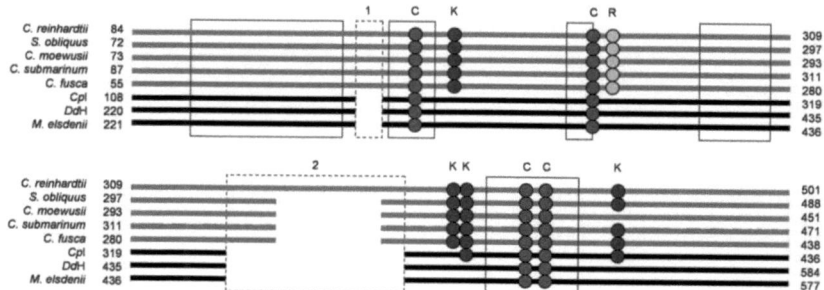

Fig. 1 Sequence alignment of the H-domain primary structure of homolog pro- and eukaryotic [FeFe] hydrogenases (black and green bars, respectively; scoring matrix BLOSUM 62). C- and N-terminal domains are trimmed for optimal fit of the alignment. Areas of high sequence similarity are marked by straight boxes. Dashed boxes show two "insertions" (1, 2) preserved in chlorophyta-type hydrogenases exclusively. According to homology models of HydA1 of *C. reinhardtii*, these sequences form a loop region replacing the F-domain of bacterial [FeFe] hydrogenases.[67] Cysteines coordinating the H-cluster (C, chestnut) are well-preserved in all [FeFe] hydrogenases. Residues K (lysine, blue) and R (arginine, grey) form a positively charged binding niche for the interaction with *in vivo* electron donor ferredoxin. This contact niche is chlorophyta-specific as well.

RESULTS

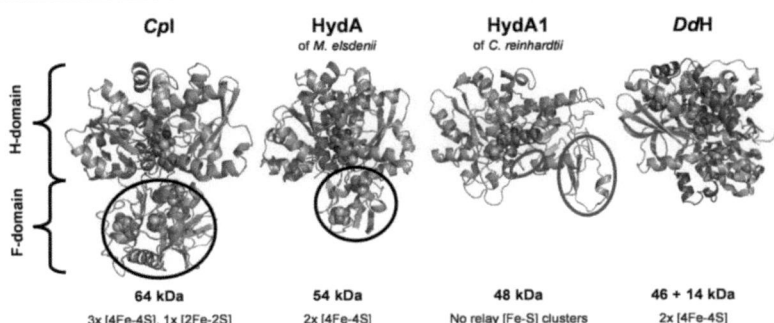

Fig. 2 Comparison of [FeFe] hydrogenases regarding the F-domain and overall structural differences. On the left site, CpI is drawn as a cartoon model from the published structure.[50] Next to CpI, only the structure of DdH (far left) was resolved by X-ray crystallography,[59] HydA and HydA1 of *M. elsdenii* and *C. reinhardtii*, respectively, have been designed by homology modelling. The F-domain and relay clusters of CpI and *M. elsdenii* HydA are marked black. For CpI, four "F-clusters" are annotated. HydA binds two [4Fe-4S] compounds. Instead of the F-domain, HydA1 exhibits an algal-specific "insertion" (red). A positively charged binding niche for interaction with ferredoxin is highlighted (blue).[67] DdH is a heterodimer with a 14 kDa chain folded around the catalytic 46 kDa catalytic subunit like a belt (red cartoon). The large subunit exhibits two [4Fe-4S] clusters wiring the H-cluster to the protein surface.

atoms are labelled "proximal" and "distal". Catalysis is thought to take place at a free binding site of the distal iron atom.[50,59,70]

Different redox states have been described for the H-cluster. The oxidized, catalytically active "H_{ox}" state is paramagnetic and EPR-active. The distal iron atom of the $[2Fe]_H$ moiety Fe_d is less reduced than the proximal iron atom Fe_p, giving the characteristic [4Fe-4S]$^{2+}$–$Fe_p(I)Fe_d(II)$ assignment.[71] One CO is found in a bridging position as identified by its typical vibrational absorption around 1800 cm^{-1} (see below). H_{ox} can bind a molecule CO at the Fe_d binding site. The paramagnetic state "H_{ox}–CO" is annotated as [4Fe-4S]$^{2+}$–$Fe_p(I)Fe_d(II)$–CO.[70,72,73] Carbon monoxide is a potent inhibitor of [FeFe] hydrogenase activity.[70,74] Furthermore, all [FeFe] hydrogenases are sensitive to oxygen inactivation, and oxygen competes with CO for the same binding site.[25,75,76] In contrast to oxygen inactivation, inhibition by CO is largely, but not entirely, reversible.[61,70,72,77] Reduction of the distal iron atom gives "H_{red}". This diamagnetic state is assigned as [4Fe-4S]$^{2+}$–$Fe_p(I)Fe_d(I)$ or hybrid species [4Fe-4S]$^{+}$–$Fe_p(II)Fe_d(II)$–H^{-}, alternatively, and not detectable by EPR spectroscopy.[62,78,79]

The [FeFe] hydrogenases of the *Desulfovibrio* genus DdH and DvH differ from typical bacterial and algal [FeFe] hydrogenases not only in structure but also regarding their insensitivity to oxygen prior a reductive activation treatment.[72,78,80] A novel state "H_{inact}"

has been characterized for aerobically isolated DdH and DvH. In this state, the [FeFe] hydrogenases of the *Desulfovibrio*-type are catalytically inactive, EPR-silent and show a typical IR spectrum, including a CO ligand in a bridging position. The Fe_d-binding site is thought to be occupied by either OH^{-} or H_2O.[50,81] By means of a reductive treatment, H_{inact} is converted to the active form H_{ox} via a state "H_{trans}". This state is transient and slightly diamagnetic due to an one-electron reduction of the [4Fe-4S] cluster.[72] It has been characterized by EPR and Fourier-transform infrared (FTIR) spectroscopy. The states H_{inact} and H_{trans} are not defined for hydrogenases like HydA1 of *C. reinhardtii* or CpI of *C. pasteurianum* which have to be isolated under strict anaerobic and reducing conditions. Here, hydrogenases (irreversibly) inactivated by oxygen are referred to as "H_{ox}air" to avoid confusion with H_{inact}. The active site composition and precise redox state of H_{ox}air remains a matter of speculation.[75]

Recently, three different algal [FeFe] hydrogenases have been examined by EPR spectroscopy. The hydrogenases from *C. reinhardtii*, *C. moewusii* and *C. submarinum* share similar g-tensors for H_{ox} and H_{ox}–CO.[54] The CO-inhibited form of *C. reinhardtii* HydA1, *e.g.*, shows the characteristic axial EPR signal with g-values of 2.052 and 2.007 (Table 2). Therefore, the electronic configuration of the H-cluster from these algal-type

Table 2 Typical g-tensors for different prokaryotic and chlorophyta-type [FeFe] hydrogenases as determined by EPR spectroscopy. The oxidized states H_{ox} and H_{ox}–CO are EPR-active due to [4Fe-4S]$^{2+}$–$Fe_p(I)Fe_d(II)$ and [4Fe-4S]$^{2+}$–$Fe_p(I)Fe_d(II)$–CO, respectively

Organism	H_{ox}	H_{ox}–CO	Reference
Clostridium acetobutylicum	n.d.	2.075, 2.009, 2.009	Von Abendroth 2008 (58)
Clostridium pasteurianum	n.d.	2.072, 2.006, 2.006	Bennet 2000 (117)
Desulfovibrio desulfuricans	2.100, 2.040, 1.999	2.065, 2.007, 2.001	Silakov 2007 (71)
Chlamydomonas reinhardtii	2.102, 2.040, 1.998	2.052, 2.007, 2.007	Kamp 2008 (54)
Chlamydomonas moewusii	2.103, 2.038, 1.998	2.052, 2.008, 2.008	Kamp 2008 (54)
Chlorococcum submarinum	2.100, 2.040, 1.998	2.056, 2.008, 2.008	Kamp 2008 (54)

RESULTS

[FeFe] hydrogenases seems to be similar. Also, it accordingly exhibits similarities to the active sites of the bacterial [FeFe] hydrogenases thus far examined. Still, distinct differences to prokaryotic hydrogenases suggest a slightly different electronic structure of the H-cluster in comparison to DdH which has been characterized by EPR spectroscopy in greater detail before.[24,71,77] Table 2 summarizes the EPR characteristics for some relevant [FeFe] hydrogenases.

Configuration of the H-cluster

The diamagnetic H_{red} state is not accessible by EPR spectroscopy. However, using X-ray absorption spectroscopy (XAS) and FTIR spectroscopy it is possible to get a picture of the H-cluster independent of the redox state. XAS at the K-edge of iron in particular is possible only with chlorophyta-type [FeFe] hydrogenases. The signals from accessory [Fe-S] clusters hamper this iron-specific analysis in bacterial hydrogenases.

By extended X-ray absorption fine structure (EXAFS, a special XAS technique that takes in account the *extended* reach of the absorption edge), the coordination of the iron atoms in [2Fe]$_H$ and [4Fe-4S] cluster could have been distinguished for *C. reinhardtii* HydA1.[69] EXAFS on the H_{red} form of the H-cluster confirmed an overall geometry similar to that of bacterial hydrogenases. The H-cluster remains essentially unperturbed upon hydrogen gas treatment, but oxidation with CO (giving H_{ox}–CO) revealed an increased number of CO ligands at the [2Fe]$_H$ moiety and a –0.1 Å elongation of the Fe$_p$(I)–Fe$_d$(II) distance.[69] Although EXAFS can not directly detect electronic states, this elongation is easily attributable to a Fe$_p$(I)–Fe$_d$(I) attraction in H_{red} lifted upon oxidation.[70] In bacterial [FeFe] hydrogenases, formation of a bridging CO between the [2Fe]$_H$ iron atoms was observed as a consequence of oxidative treatment.[25,70,71,81] The intensified attribution of CO ligands in H_{ox}–CO can be explained alike. Fig. 3 shows the EXAFS analysis in a structural model for the oxidized H_{ox}–CO H-cluster.

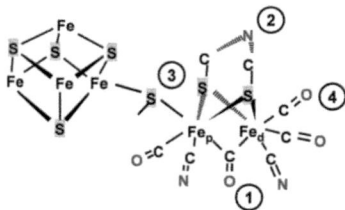

Fig. 3 Structure of the *C. reinhardtii* HydA1 H-cluster in its H_{ox}–CO state as derived from EXAFS. The distance of Fe$_p$ and Fe$_d$ is 2.62 Å, about 0.1 Å longer than reported for H_{red}.[69] One CO is found in a bridging position (1). The heteroatom in the dithiolate ligand (2) was *not* resolved by EXAFS. We follow a recent EPR study[116] and display the ligand as an azadithiolate bridge. A cysteine residue (3) binds the catalytic di-iron unit to the [4Fe-4S] cluster. In H_{ox}–CO, extrinsic CO (4) occupies the Fe$_d$ binding site.

The CO and CN$^-$ ligands of the H-cluster are uncommon in nature, due to the high reactivity of most notably CN$^-$. Maturation of [FeFe] hydrogenases and the *in vivo* formation of the H-cluster in particular is a field of active research.[57,82,83] Several publications report the *in situ* synthesis is H-cluster analogues, mimicking the unique ligand substitution.[84-87] However, infrared spectroscopy allows for the investigation on these specific groups and the actual situation of the H-cluster in consequence. With an absorption in the range of 2100 to 1800 cm^{-1}, the vibrational modes of CO and CN$^-$ ligands can be analyzed by FTIR spectroscopy without interference from the protein backbone. The typical IR spectrum of the H-cluster can be subdivided into three main regions. Absorption from 2100 to 2050 cm^{-1} is attributable to CN$^-$ stretching vibrations. Within 2050 to 1820 cm^{-1}, the different vibrational modes of the CO ligands absorb incoming IR radiance. This region is the most complex part of the spectrum. From 1820 to 1790 cm^{-1} approximately, the bridging CO can be detected.[72,73,88,89]

The [FeFe] hydrogenase HydA1 from *C. reinhardtii* was subject of a spectro-electrochemical analysis. The enzyme was investigated by FTIR spectroscopy in a "Moss cell".[90] By application of a certain voltage, the redox states of the enzyme can be adjusted without any additional chemical treatment. HydA1 was found to exhibit typical bands in the spectrum from 2100 to 1800 cm^{-1}. The H_{ox} state can be recognized from a prominent absorption band at 1940 cm^{-1} (stretch frequency Fe$_d$-CO) and a typical 1800 cm^{-1} peak due to the bridging CO stretch frequency.[72,91,92] Lowering the potential, bands at 1935 cm^{-1} and 1891 cm^{-1} (stretch frequency Fe$_p$-CO) emerge which are attributable to the H_{red} state.[90] Interestingly, a band around 1800 cm^{-1} is observed which might indicate that the CO bridge in HydA1 is not lifted upon reduction of the active site. This observation has important implications as it argues against a catalytic reaction mechanism that involves both [2Fe]$_H$ iron atoms and a bridging hydride.[93] Note that a bridged H_{red} state is in contradiction to what has been reasoned from EXAFS for HydA1 of *C. reinhardtii*.[69] Below –500 mV vs SHE, HydA1 was found to adopt a "super reduced" state, comparable to what Albracht *et al.* observed with the bacterial DdH [FeFe] hydrogenase.[72] Potentials more positive than –100 mV gave H_{ox}–CO due to "cannibalization"–an effect indicative of protein degradation and subsequent release of CO which binds H-clusters still intact.[72,77]

Electrochemical analysis of HydA1 of *C. reinhardtii*

In the spectro-electrochemical studies on HydA1 of *C. reinhardtii*, voltage was applied to a solution of protein to adjust for the redox state of the H-cluster.[90] Protein film electrochemistry analyzes protein (mono-) layers in contact with a conductive surface.[23,94,95] Current is recorded as a function of the applied potential and is equivalent to the catalytic redox activity of the protein layer. At potentials more negative than the redox potential of the bound protein, electrons are driven from the (working) electrode to the enzyme, hence reducing immobilized enzyme. Potential values more positive result in an oxidation of the protein layer. Working electrodes are commonly made of gold, platinum and different kinds of graphite. Proteins usually prefer binding to graphite,[23,96] and metal surfaces need to be modified by mercaptoterminated hydrocarbon molecules to circumvent protein degradation and background current due to surface oxidation and absorbed hydrogen layers.[97,98] Modified noble metal electrodes present tailor-made binding surfaces and, in case of gold, provide

RESULTS

the possibility for concerted spectro-electrochemical analyses.[99,100] Graphite electrodes guarantee fast and rather unspecific binding.

Just recently, the [FeFe] hydrogenase HydA1 of *C. reinhardtii* was shown to be catalytically active immobilized on a modified gold electrode.[98] HydA1 was bound to a rough gold surface *via* two different carboxy-terminated self-assembled monolayers (SAM). Current and hydrogen evolution was recorded after immobilization of the hydrogenase and addition of methylviologen as electron shuttle. Whether the SAM was formed from mercaptopropionic acid (3C) or mercaptoundecanoic acid (11C), direct electron transfer (DET) from the electrode surface to the hydrogenase has *not* been observed.[98] By Surface Enhanced Infrared Spectroscopy (SEIRAS), binding kinetics were recorded, and *via* Surface Plasmon Resonance (SPR), the amount of bound protein could have been determined. SEIRAS is an IR spectroscopic technique which enhances the vibrational absorption of adsorbed molecules by more than two orders of magnitude.[101-103] This is due to plasmon excitation in metal surfaces (Au, Pt, Pd) by an incident electric field, an effect utilized in Raman spectroscopy as well. The novel set-up can serve as a device for electrochemical hydrogen production at defined specific activities. Furthermore, IR spectro-electrochemical investigations are possible which bring forth the advantage of full control of the protein layer redox activity *via* potential.

Armstrong *et al.* established protein film electrochemistry on pyrolytic graphite edge for many [FeFe] and [NiFe] hydrogenases.[23] However, immobilization of a chlorophyta-type hydrogenase has not been reported up to now. In recent studies, it was shown that HydA1 of *C. reinhardtii* directly exchanges electrons with the pyrolytic graphite edge electrode.[61,75] This is not trivial as electrons need to tunnel directly into the active site due to the missing [Fe-S] cluster wire in algal hydrogenases. For the first time, the bidirectional character of HydA1 was shown. Fig. 4

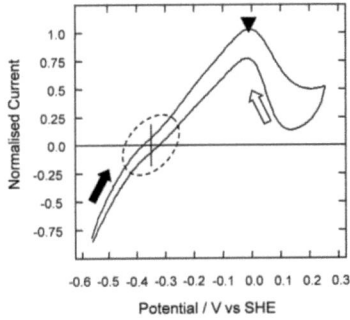

Fig. 4 Cyclic voltammogram of *C. reinhardtii* HydA1 immobilized on a pyrolytic graphite edge rotating disc electrode. At −500 mV *vs* SHE, reductive current is the same as oxidative current recorded at −200 mV. The midpoint redox potential is about −350 mV *vs* SHE. An area of inflection is marked by the dashed oval. At potentials more positive than 0 mV (▼), HydA1 loses activity due to anaerobic inactivation.[60] The arrows give direction of forward (solid) and backward scan (open). Experimental conditions: 100 mM KPi buffer pH 6.0, 20 °C, 1 bar H$_2$, electrode rotation rate 3000 rpm, scan rate 20 mV/s.

displays a cyclic voltammogram of the *C. reinhardtii* hydrogenase. At a given overpotential with regard to the redox potential in either the reduction or oxidation direction, the magnitude of the reduction current is similar to that of the oxidation current. The enzyme exhibits approximately similar activities in reduction (−500 mV *vs* SHE) and oxidation (−200 mV *vs* SHE) at pH 6.0. The inflection marked by the dashed oval reflects the bit of extra driving force necessary due to the lack of accessory clusters in HydA1. Recorded under an atmosphere of 100% hydrogen, it is interesting to note that proton-reducing (hydrogen evolution) activity was not hampered by product (hydrogen) inhibition. A process, commonly referred to as "anaerobic inactivation", occurs at potentials more positive than 0 mV. Current drops and is recovered on the back scan at appropriate rate.

Inactivation at positive potential values is reversible to a different extent depending which enzyme is probed.[60] A new state "H$_{ox}$inact" is defined for [FeFe] hydrogenases under this conditions,[104,105] setting anaerobic inactivation apart from H$_{inact}$ and H$_{ox}$air.[72,78,80]

Fully reversible inhibition of hydrogen oxidation by CO was shown for HydA1 of *C. reinhardtii*, alongside protection of the H-cluster by CO against oxygen damage. Surprisingly, reaction with oxygen was found to be ten times slower than that reported for the bacterial-type [FeFe] hydrogenase *Dd*H.[60,75] The irreversible oxygen inactivation of HydA1 is further slowed upon tenfold excess of hydrogen, due to competition for the active site. In summary, experimental evidence is demonstrated that hydrogen, oxygen and CO bind to the H-cluster at the same site. Presumably, this is the distal iron atom of the [2Fe]$_H$ moiety.[50,59,75] Note that a single binding site is not obvious for a [6Fe-6S] compound or di-iron reaction centre, at least.

From a recent EXAFS analysis, it was observed that the [4Fe-4S] part of the H-cluster is disrupted exclusively upon oxygen inactivation. The catalytically active [2Fe]$_H$ unit is initially left intact.[75] As CO is not known to bind to cubane clusters, it must bind to the [2Fe]$_H$ moiety. Thus, the electrochemical demonstration that CO protects the active site from oxygen indicates that oxygen does not *directly* attack the cubane cluster.[70,74,106] Taking these independent observations into account, two ideas of how oxygen inactivates the H-cluster present themselves. Oxygen is either reduced to a reactive oxygen species (*e.g.*, superoxide) or takes one electron from the [4Fe-4S] cluster *via* through-bond oxidation. Reactive oxygen might then be able to attack the cubane subcluster directly.[75,107] Both effects, however, result in oxidation of the [4Fe-4S] cluster and subsequent loss of iron. The Fe K-edge of oxygen-treated samples of *C. reinhardtii* HydA1 displayed a huge peak indicative of ferrous Fe^{2+}.[69] Oxidative disassembly of [Fe-S] clusters is a frequently observed phenomenon (see ref. 103). For the first time, oxygen inactivation was followed by protein film electrochemistry and EXAFS. Due to the relatively slow reaction with oxygen and the absence of any other [Fe-S] compounds than the H-cluster, HydA1 is the only [FeFe] hydrogenase suitable for the set of experiments chosen here.

The interaction of HydA1 with ferredoxin PetF of *C. reinhardtii*

In green algae, hydrogen production is light dependent and coupled to the photosynthetic transport chain *via* ferredoxin

PetF. Although six *fdx* genes were discovered in *C. reinhardtii*, only PetF ("Photosynthetic electron transfer Ferredoxin") is able to reduce the hydrogenase *in vitro*.[108] PetF is discussed to be the central branching point of reducing power in sulfur-deprived algae.[46,108,109] Thus, HydA1 and ferredoxin-NADPH-reductase, which both use ferredoxin as an electron donor, compete for electrons of the photosynthetic transport chain at the level of PetF. It has been shown that this competition determines the hydrogen evolution capacities of the algal cell.[14,110]

A recent study examines the interaction of *C. reinhardtii* proteins HydA1 and PetF with the help of site directed mutagenesis.[67] Several variants were specifically designed on the basis of predicted electrostatic surface distribution and prior *in silico* docking analyses and have been generated using the overexpression system described above.[55,58] Mapping the Michaelis-Menten kinetics of several variants of HydA1 and PetF *via* methylviologen and PetF reduction, a ferredoxin-specific effect was observed for especially two lysine residues. The electron surface potential of HydA1 was simulated to become more negative in these variants. In non-conservative variants, V_{max} is lowered to 60% and 10%, respectively, while hydrogen evolution activity is unchanged for methylviologen as electron donor. These analyses in combination with *in silico* docking studies show that electrostatic interactions between the lysine residues and the C-terminus of PetF play a major role in complex formation and electron transfer.[67] Mapping of significant *C. reinhardtii* HydA1 and PetF residues represents an important method for controlling the physiological photosynthetic electron flow in favour of light-driven hydrogen production.

Outlook

Green algae of the chlorophyta-type encode for [FeFe] hydrogenases smaller and more simple than those known from bacteria. While prokaryotic [FeFe] hydrogenases use a wire of two to four [Fe-S] clusters for translocation of electrons to the active site H-cluster, the algal hydrogenases lacks this accessory subdomain. Therefore, HydA1 of *C. reinhardtii* represents a "minimal catalyst for biological hydrogen production".[69] In this review, we have given a brief overview on the history of an interesting class of [Fe-S] enzymes, the [FeFe] hydrogenases of green algae. We report the most recent biophysical characterizations by electron spin resonance, Fourier-transform infrared spectroscopy, X-ray absorption spectroscopy and protein film electrochemistry. Furthermore, we summarize a work analyzing the specific HydA1–PetF interaction crucial in *C. reinhardtii* photobiological hydrogen production.

Many aspects of the algal hydrogen turnover are still unclear and deserve intensive research. In particular, protein biosynthesis and maturation of the H-cluster is a matter of debate. Organisms encoding for a [FeFe] hydrogenase need at least three maturation enzymes (HydE, HydF and HydG) that catalyze the ligation of the $[2Fe]_H$ moiety and translocation of the prosthetic group onto the hydrogenase apo-protein.

In *C. reinhardtii*, HydE and HydF form a single-chain protein complex.[57,82] HydF is thought to act as the central "scaffold" protein, a sort of construction site from where the H-cluster is transferred to the apoprotein, presumably with the help of HydF GTPase activity.[83,111] Open questions include the specific part of HydE, HydF and HydG in *in vivo* maturation as well as the origins of the CO and CN⁻ ligands. While there is some data suggesting the origin of the CN⁻ ligands in [NiFe] hydrogenases,[112,113] the precursors of the ligand groups in [FeFe] hydrogenases have not been identified yet.

The interest in exploitation of algal hydrogenases mainly results from their role in photobiological hydrogen production. Many studies report on the need to produce renewable "biohydrogen" by the use of sunlight and hydrogenases-catalyzed electrolysis.[31,39] One approach is to immobilize both PSI and PSII on electrically linked gold electrodes.[114] On the anodic site, PSII is bound to a special carbohydrate polymer which has been shown to work best for large protein complexes.[115] Water is split when the cell is illuminated, and electrons travel *via* the gold surface to the connected PSI electrode. In analogy to the photosynthetic electron transfer chain, electrons are excited by light at PSI a second and actively transferred from PSI to the hydrogenase.

In this "hydrogen battery", anode and cathode are separated in two gas-sealed cells. Charge exchange is ensured by electron coupling of PSII and PSI electrodes. Protons as the product of water oxidation and the substrate of hydrogen production are free to diffuse from the anode to cathode compartments. The photobiological hydrogen device produces current, oxygen and hydrogen upon illumination. This setup, on the one hand, allows for screening of the optimal components. Each enzyme module can be exchanged by a likely protein—in case of PSII, a stable D1 variant is of interest, *e.g.* from a thermophilic organisms.[114] On the other hand, the battery can directly serve as a fuel cell once all components have been optimized.

All together the new insights into the structural properties of the algal hydrogenases might be used to enhance the photosynthetic hydrogen production process in unicellular green algae and help unravel the molecular principals of hydrogen turnover.

Acknowledgements

First off, we would like to thank Gabrielle Goldet for her critical revision of the manuscript. We would also like to acknowledge our co-workers: Anastasios Melis (Berkeley); Michael Haumann (Berlin); Henning Krassen and Joachim Heberle (Bielefeld); Alexey Silakov, Ed Reijerse and Wolfgang Lubitz (Mülheim); Gabrielle Goldet, Kylie Vincent and Fraser Armstrong (Oxford); Laurence Girbal and Philippe Soucaille (Toulouse); Lore Florin, Anja Hemschemeier, Annette Kaminski, Christina Kamp, Anestis Tsokoglou, Gregory von Abendroth and Martin Winkler (Bonn and Bochum) for their invaluable contributions in the analyses of algal hydrogenases, especially the [FeFe] hydrogenase HydA1 of *Chlamydomonas reinhardtii*. The research was supported by grants from the Deutsche Forschungsgemeinschaft (SFB-480) and the EU/Energy Network SolarH2 (FP7 contract 212508).

References

1 M. Stephenson and L. H. Stickland, *Biochem. J.*, 1931, **25**, 205–14.
2 H. Gaffron, *Nature*, 1939, **143**, 204–5.
3 H. Gaffron, *Science*, 1940, **91**, 529–30.
4 H. Gaffron and J. Rubin, *J. Gen. Physiol.*, 1942, **26**, 219–40.
5 T. S. Stuart and H. Gaffron, *Plant Physiol.*, 1972, **50**, 136–40.
6 A. Melis, L. Zhang, M. Forestier, M. L. Ghirardi and M. Seibert, *Plant Physiol.*, 2000, **122**, 127–36.
7 D. D. Wykoff, J. P. Davies, A. Melis and A. R. Grossman, *Plant Physiol.*, 1998, **117**, 129–39.
8 L. Zhang, T. Happe and A. Melis, *Planta*, 2002, **214**, 552–61.

RESULTS

9 S. Fouchard, A. Hemschemeier, A. Caruana, J. Pruvost, J. Legrand, T. Happe, G. Peltier and L. Cournac, *Appl. Environ. Microbiol.*, 2005, **71**, 6199–205.
10 A. Hemschemeier and T. Happe, *Biochem. Soc. Trans.*, 2005, **33**, 39–41.
11 A. Melis and T. Happe, *Plant Physiol.*, 2001, **127**, 740–8.
12 P. Decottignies, V. Flesch, C. Gerard-Hirne and P. Le Marechal, *Plant Physiol. Biochem.*, 2003, **41**, 637–42.
13 D. B. Knaff, in *Oxygenic Photosynthesis: The Light Reactions*, ed. D. R. Ort and D. F. Yocum, Kluwer Academic Publishers, Dordrecht, 1996, pp. 333–61.
14 K. Hemschemeier, S. Fouchard, L. Cournac, G. Peltier and T. Happe, *Planta*, 2007, **227**, 397–407.
15 M. Gibbs, R. P. Gfeller and C. Chen, *Plant Physiol.*, 1986, **82**, 160–6.
16 R. P. Gfeller and M. Gibbs, *Plant Physiol.*, 1985, **77**, 509–11.
17 T. Happe, A. Hemschemeier, M. Winkler and A. Kaminski, *Trends Plant Sci.*, 2002, **7**, 246–50.
18 A. Hemschemeier, A. Melis and T. Happe, *Photosynth. Res.*, 2009, DOI: 10.1007/s11120-009-9415-5.
19 M. Winkler, A. Hemschemeier, C. Gotor, A. Melis and T. Happe, *IJHE*, 2002, **27**, 1431–9.
20 P. M. Vignais and B. Billoud, *Chem. Rev.*, 2007, **107**, 4206–72.
21 D. S. Horner, B. Heil, T. Happe and T. M. Embley, *Trends Biochem. Sci.*, 2002, **27**, 148–53.
22 M. Cammack, in *Hydrogen as a Fuel - Learning from Nature*, ed. R. Cammack, M. Frey and R. Robson, Taylor & Francis, London, New York, 2001, pp. 1–9.
23 K. A. Vincent, A. Parkin and F. A. Armstrong, *Chem. Rev.*, 2007, **107**, 4366–413.
24 W. Lubitz, E. Reijerse and M. van Gastel, *Chem. Rev.*, 2007, **107**, 4331–65.
25 M. W. W. Adams, *Biochim. Biophys. Acta, Bioenerg.*, 1990, **1020**, 115–45.
26 R. Cammack, *Nature*, 1999, **397**, 214–5.
27 S. P. Albracht, *Biochim. Biophys. Acta, Bioenerg.*, 1994, **1188**, 167–204.
28 R. K. Thauer, A. R. Klein and G. C. Hartmann, *Chem. Rev.*, 1996, **96**, 3031–42.
29 S. Shima, O. Pilak, S. Vogt, M. Schick, M. S. Stagni, W. Meyer-Klaucke, E. Warkentin, R. K. Thauer and U. Ermler, *Science*, 2008, **321**, 572–5.
30 M. Frey, *ChemBioChem*, 2002, **3**, 153–60.
31 M. L. Ghirardi, A. Dubini, J. Yu and P. C. Maness, *Chem. Soc. Rev.*, 2009, **38**, 52–61.
32 M. W. Adams and E. I. Stiefel, *Curr. Opin. Chem. Biol.*, 2000, **4**, 214–20.
33 A. Silakov, P. Oluwole, O. Troshina, P. Lindblad, E. Leitao, P. Oliveira and P. Tamagnini, *Planta*, 2004, **218**, 350–9.
34 L. C. Sun, B. Akermark and S. Ott, *Coord. Chem. Rev.*, 2005, **249**, 1653–1663.
35 C. Tard, X. Liu, S. K. Ibrahim, M. Bruschi, L. De Gioia, S. C. Davies, X. Yang, L. S. Wang, G. Sawers and C. J. Pickett, *Nature*, 2005, **433**, 610–3.
36 J. F. Capon, S. Ezzaher, F. Gloaguen, F. Y. Petillon, P. Schollhammer and J. Talarmin, *Chem.–Eur. J.*, 2008, **14**, 1954–64.
37 L. Duan, M. Wang, P. Li, Y. Na, N. Wang and L. Sun, *Dalton Trans.*, 2007, 1277–83.
38 G. A. Felton, A. K. Vannucci, J. Chen, L. T. Lockett, N. Okumura, B. J. Petro, U. I. Zakai, D. H. Evans, R. S. Glass and D. L. Lichtenberger, *J. Am. Chem. Soc.*, 2007, **129**, 12521–30.
39 M. Wang, Y. Na, M. Gorlov and L. Sun, *Dalton Trans.*, 2009, 6458–67.
40 A. Melis, M. Seibert and T. Happe, *Photosynth. Res.*, 2004, **82**, 277–88.
41 A. Melis and T. Happe, *Photosynth. Res.*, 2004, **80**, 401–9.
42 E. Kessler, *Arch. Microbiol.*, 1973, **93**, 91–100.
43 F. B. Abeles, *Plant Physiol.*, 1964, **39**, 169–76.
44 T. Happe and J. D. Naber, *Eur. J. Biochem.*, 1993, **214**, 475–81.
45 P. G. Roessler and S. Lien, *Plant Physiol.*, 1984, **75**, 705–9.
46 P. G. Roessler and S. Lien, *Plant Physiol.*, 1984, **76**, 1086–9.
47 T. Happe, B. Mosler and J. D. Naber, *Eur. J. Biochem.*, 1994, **222**, 769–74.
48 J. Schnackenberg, R. Schulz and H. Senger, *FEBS Lett.*, 1993, **327**, 21–4.
49 T. Happe and A. Kaminski, *Eur. J. Biochem.*, 2002, **269**, 1022–32.
50 J. W. Peters, W. N. Lanzilotta, B. J. Lemon and L. C. Seefeldt, *Science*, 1998, **282**, 1853–8.
51 L. Florin, A. Tsokoglou and T. Happe, *J. Biol. Chem.*, 2001, **276**, 6125–32.
52 M. Winkler, B. Heil and T. Happe, *Biochim. Biophys. Acta, Gene Struct. Expression*, 2002, **1576**, 330–4.
53 P. M. Vignais, B. Billoud and J. Meyer, *FEMS Microbiol. Rev.*, 2001, **25**, 455–501.
54 C. Kamp, A. Silakov, M. Winkler, E. J. Reijerse, W. Lubitz and T. Happe, *Biochim. Biophys. Acta, Bioenerg.*, 2008, **1777**, 410–6.
55 L. Girbal, G. von Abendroth, M. Winkler, P. M. Benton, I. Meynial-Salles, C. Croux, J. W. Peters, T. Happe and P. Soucaille, *Appl. Environ. Microbiol.*, 2005, **71**, 2777–81.
56 K. Sybirna, T. Antoine, P. Lindberg, V. Fourmond, M. Rousset, V. Mejean and H. Bottin, *BMC Biotechnol.*, 2008, **8**, 73.
57 M. C. Posewitz, P. W. King, S. L. Smolinski, L. Zhang, M. Seibert and M. L. Ghirardi, *J. Biol. Chem.*, 2004, **279**, 25711–20.
58 G. von Abendroth, S. T. Stripp, A. Silakov, C. Croux, P. Soucaille, L. Girbal and T. Happe, *IJHE*, 2008, **33**, 6076–81.
59 Y. Nicolet, C. Piras, P. Legrand, C. E. Hatchikian and J. C. Fontecilla-Camps, *Structure*, 1999, **7**, 13–23.
60 K. A. Vincent, A. Parkin, O. Lenz, S. P. Albracht, J. C. Fontecilla-Camps, R. Cammack, B. Friedrich and F. A. Armstrong, *J. Am. Chem. Soc.*, 2005, **127**, 18179–89.
61 G. Goldet, C. Brandmayr, S. T. Stripp, T. Happe, C. Cavazza, J. Fontecilla-Camps and F. A. Armstrong, *J. Am. Chem. Soc.*, 2009, **131**, 14979–89.
62 E. C. Hatchikian, N. Forget, V. M. Fernandez, R. Williams and R. Cammack, *Eur. J. Biochem.*, 1992, **209**, 357–65.
63 M. Winkler, C. Maeurer, A. Hemschemeier and T. Happe, in *Biohydrogen III - Renewable Energy System by Biological Solar Energy Conversion*, ed. J. Miyake, Y. Igarashi and M. Roegner, Elsevier, Oxford, 1st edn, 2004, pp. 103–15.
64 O. Lenz, M. Bernhard, T. Buhrke, E. Schwartz and B. Friedrich, *J. Mol. Microbiol. Biotechnol.*, 2002, **4**, 255–62.
65 K. Muellner and T. Happe, *International Journal of Energy Technology and Policy*, 2007, **5**, 290–5.
66 M. Forestier, P. King, L. Zhang, M. Posewitz, S. Schwarzer, T. Happe, M. L. Ghirardi and M. Seibert, *Eur. J. Biochem.*, 2003, **270**, 2750–8.
67 M. Winkler, S. Kuhlgert, M. Hippler and T. Happe, *J. Biol. Chem.*, 2009, DOI: 10.1074/jbc.M109.053496.
68 M. Atta and J. Meyer, *Biochim. Biophys. Acta*, 2000, **1476**, 368–71.
69 S. T. Stripp, O. Sanganas, T. Happe and M. Haumann, *Biochemistry*, 2009, **48**, 5042–9.
70 B. J. Lemon and J. W. Peters, *Biochemistry*, 1999, **38**, 12969–73.
71 A. Silakov, E. J. Reijerse, S. P. Albracht, E. C. Hatchikian and W. Lubitz, *J. Am. Chem. Soc.*, 2007, **129**, 11447–58.
72 W. Roseboom, A. L. De Lacey, V. M. Fernandez, E. C. Hatchikian and S. P. Albracht, *JBIC, J. Biol. Inorg. Chem.*, 2006, **11**, 102–18.
73 A. L. De Lacey, C. Stadler, C. Cavazza, E. C. Hatchikian and V. M. Fernandez, *J. Am. Chem. Soc.*, 2000, **122**, 11232–3.
74 D. L. Erbes, D. King and M. Gibbs, *Plant Physiol.*, 1979, **63**, 1138–42.
75 S. T. Stripp, G. Goldet, C. Brandmayr, O. Saganas, K. A. Vincent, M. Haumann, F. A. Armstrong and T. Happe, *PNAS*, 2009, **106**, 17331–6.
76 A. L. De Lacey, V. M. Fernandez, M. Rousset and R. Cammack, *Chem. Rev.*, 2007, **107**, 4304–30.
77 S. P. Albracht, W. Roseboom and E. C. Hatchikian, *JBIC, J. Biol. Inorg. Chem.*, 2006, **11**, 88–101.
78 D. S. Patil, J. J. Moura, S. H. He, M. Teixeira, B. C. Prickril, D. V. DerVartanian, H. D. Peck, Jr., J. LeGall and B. H. Huynh, *J. Biol. Chem.*, 1988, **263**, 18732–8.
79 A. J. Pierik, W. R. Hagen, J. S. Redeker, R. B. Wolbert, M. Boersma, M. F. Verhagen, H. J. Grande, C. Veeger, P. H. Mutsaers and R. H. Sands, *Eur. J. Biochem.*, 1992, **209**, 63–72.
80 H. M. Van Der Westen, S. G. Mayhew and C. Veeger, *FEBS Lett.*, 1978, **86**, 122–6.
81 Y. Nicolet, A. L. de Lacey, V. Vernede, V. M. Fernandez, E. C. Hatchikian and J. C. Fontecilla-Camps, *J. Am. Chem. Soc.*, 2001, **123**, 1596–601.
82 M. C. Posewitz, P. W. King, S. L. Smolinski, R. D. Smith, A. R. Ginley, M. L. Ghirardi and M. Seibert, *Biochem. Soc. Trans.*, 2005, **33**, 102–4.
83 D. W. Mulder, D. O. Ortillo, D. J. Gardenghi, A. V. Naumov, S. S. Ruebush, R. K. Szilagyi, B. Huynh, J. B. Broderick and J. W. Peters, *Biochemistry*, 2009, **48**, 6240–8.
84 R. Mejia-Rodriguez, D. Chong, J. H. Reibenspies, M. P. Soriaga and M. Y. Darensbourg, *J. Am. Chem. Soc.*, 2004, **126**, 12004–14.

RESULTS

85 X. Hu, B. S. Brunschwig and J. C. Peters, *J. Am. Chem. Soc.*, 2007, **129**, 8988–98.
86 D. E. Schwab, C. Tard, E. Brecht, J. W. Peters, C. J. Pickett and R. K. Szilagyi, *Chem. Commun.*, 2006, 3696–8.
87 M. H. Cheah, C. Tard, S. J. Borg, X. Liu, S. K. Ibrahim, C. J. Pickett and S. P. Best, *J. Am. Chem. Soc.*, 2007, **129**, 11085–92.
88 K. Nakamoto, in *Infrared and Raman Spectra of Inorganic and Coordination Compounds*, ed. K. Nakamoto, John Wiley & Sons, New York, 1997, pp. 53–6.
89 A. J. Pierik, M. Hulstein, W. R. Hagen and S. P. Albracht, *Eur. J. Biochem.*, 1998, **258**, 572–8.
90 A. Silakov, C. Kamp, E. Reijerse, T. Happe and W. Lubitz, *Biochemistry*, 2009, **48**, 7780–6.
91 Z. Chen, B. J. Lemon, S. Huang, D. J. Swartz, J. W. Peters and K. A. Bagley, *Biochemistry*, 2002, **41**, 2036–43.
92 T. M. Van Der Spek, A. F. Arendsen, R. P. Happe, S. Yun, K. A. Bagley, D. J. Stufkens, W. R. Hagen and S. P. Albracht, *Eur. J. Biochem.*, 1996, **237**, 629–34.
93 S. Löscher, L. Schwartz, M. Stein, S. Ott and M. Haumann, *Inorg. Chem.*, 2007, **46**, 11094–105.
94 S. J. Elliott, C. Leger, H. R. Pershad, J. Hirst, K. Heffron, N. Ginet, F. Blasco, R. A. Rothery, J. H. Weiner and F. A. Armstrong, *Biochim. Biophys. Acta, Bioenerg.*, 2002, **1555**, 54–9.
95 C. Léger, S. J. Elliott, K. R. Hoke, L. J. Jeuken, A. K. Jones and F. A. Armstrong, *Biochemistry*, 2003, **42**, 8653–62.
96 M. Hambourger, M. Gervaldo, D. Svedruzic, P. W. King, D. Gust, M. Ghirardi, A. L. Moore and T. A. Moore, *J. Am. Chem. Soc.*, 2008, **130**, 2015–22.
97 J. Wang, in *Analytical Electrochemistry*, ed. J. Wang, Wiley VCH, Hoboken, 3rd edn, 2006, pp. 136–8.
98 H. Krassen, S. T. Stripp, G. von Abendroth, K. Ataka, T. Happe and J. Heberle, *J. Biotechnol.*, 2009, **142**, 3–9.
99 K. Ataka and J. Heberle, *Biochem. Soc. Trans.*, 2008, **36**, 986–91.
100 K. Ataka and J. Heberle, *J. Am. Chem. Soc.*, 2004, **126**, 9445–57.
101 M. Osawa, in *Handbook of Vibrational Spectroscopy*, ed. J. M. Chalmers and P. R. Griffiths, Wiley, Chichester, 2002, pp. 785–99.
102 M. Osawa, *Bull. Chem. Soc. Jpn.*, 1997, **70**, 2861–80.
103 K. Ataka and J. Heberle, *Anal. Bioanal. Chem.*, 2007, **388**, 47–54.
104 A. Parkin, C. Cavazza, J. C. Fontecilla-Camps and F. A. Armstrong, *J. Am. Chem. Soc.*, 2006, **128**, 16808–15.
105 A. K. Jones, S. E. Lamle, H. R. Pershad, K. A. Vincent, S. P. Albracht and F. A. Armstrong, *J. Am. Chem. Soc.*, 2003, **125**, 8505–14.
106 G. J. Kubas, *Chem. Rev.*, 2007, **107**, 4152–205.
107 J. C. Crack, J. Green, M. R. Cheesman, N. E. Le Brun and A. J. Thomson, *Proc. Natl. Acad. Sci. U. S. A.*, 2007, **104**, 2092–7.
108 J. Jacobs, S. Pudollek, A. Hemschemeier and T. Happe, *FEBS Lett.*, 2009, **583**, 325–9.
109 J. M. Schmitter, J. P. Jacquot, F. de Lamotte-Guery, C. Beauvallet, S. Dutka, P. Gadal and P. Decottignies, *Eur. J. Biochem.*, 1988, **172**, 405–12.
110 T. Rühle, A. Hemschemeier, A. Melis and T. Happe, *BMC Plant Biol.*, 2008, **8**, 107.
111 S. E. McGlynn, S. S. Ruebush, A. Naumov, L. E. Nagy, A. Dubini, P. W. King, J. B. Broderick, M. C. Posewitz and J. W. Peters, *JBIC, J. Biol. Inorg. Chem.*, 2007, **12**, 443–7.
112 A. Bock, P. W. King, M. Blokesch and M. C. Posewitz, *Adv. Microbial Physiol.*, 2006, **51**, 1–71.
113 L. Forzi and R. G. Sawers, *BioMetals*, 2007, **20**, 565–78.
114 B. Esper, A. Badura and M. Rogner, *Trends Plant Sci.*, 2006, **11**, 543–9.
115 A. Badura, B. Esper, K. Ataka, C. Grunwald, C. Woll, J. Kuhlmann, J. Heberle and M. Rogner, *Photochem. Photobiol.*, 2006, **82**, 1385–90.
116 A. Silakov, B. Wenk, E. Reijerse and W. Lubitz, *Phys. Chem. Chem. Phys.*, 2009, **11**, 6592–9.
117 B. Bennett, B. J. Lemon and J. W. Peters, *Biochemistry*, 2000, **39**, 7455–60.

DISCUSSION

DISCUSSION

3.1 Heterologous expression and synthesis of [FeFe] hydrogenases

Section (2.1) reports on recent improvements in heterologous synthesis and purification of *Cr*HydA1, the [FeFe] hydrogenase of *C. reinhardtii*. Isolation yield was increased ten-fold (up to 1.000 µg L^{-1}) by adapting the eukaryotic hydA1$_{Cr}$ gene to the *C. acetobutylicum* codon usage. Furthermore, elongation of the strep-tag was found to optimize binding to the StrepTactin column. EPR confirmed the integrity of the active site H-cluster. Furthermore, expression and protein synthesis of streptagged *Ca*HydA in its native host was established by homologous recombination. Protein yield was doubled to 800 µg L^{-1}. The V$_{max}$ for *Ca*HydA was found to be 1.750 µmol H$_2$ min^{-1} mg^{-1} which is 175-fold higher than published by Girbal et al. in 2005 [35] and in agreement with recent publications by other authors [74, 133]. The sound conditions of the H-cluster in *Ca*HydA were proven by EPR spectroscopy.

In recent years, three different systems emerged that allow for heterologous synthesis and genetic engineering of [FeFe] hydrogenases. These can be categorized according to the host organism: *C. acetobutylicum* [34, 35], *E. coli* [134, 135], and *S. oneidensis* [70]. *Clostridium acetobutylicum* is a strict anaerobe that exhibits a fermentation metabolism. *Escherichia coli* and *S. oneidensis* are facultative anaerobes that run on a "surviving" metabolism if deprived from O$_2$ [136, 137]. In Table 1 a comparison of the most important parameters can be found.

Table 1 – *Comparison of different systems for the heterologous synthesis of [FeFe] hydrogenases. Optical density (oD) as measured at 600 nm. *Specific H$_2$ evolving activity V$_{max}$ was determined with methylviologen and sodium dithionite, at 37° C and pH = 7.0. V$_{max}$ as expressed in µmol H$_2$ mg^{-1} min^{-1}. **%Vmax is H$_2$ evolving activity relative to the activity of native CrHydA1 [60].*

host organism	genetic handling	pro-moter	abs. yield [µg L^{-1}]	rel. yield (oD = 1)	max. oD	V$_{max}$*	% V$_{max}$**	reference
C. acetobutylicum	–	*thl*	1.000	250	4	625	67	[34, 35]
E. coli	+	T7	1.000	2.000	0.5	150	16	[134, 135]
S. oneidensis	++	*lac*	500	500	1	700	75	[70]

All three systems display specific advantages and drawbacks. The closely related proteobacteria *E. coli* and *S. oneidensis* are genetically well-characterised. A wide variety of protocols for transformation and synthesis of recombinant proteins can be found. Moreover, both micro organisms display great flexibility when it comes to codon usage. *Shewanella oneidensis* is suitable for the translation of eukaryotic genes in particular [70]. Electro-transformation and establishment of recombinant genetic material in *C. acetobutylicum* [138] is far less efficient. Codon usage adaptation was proven necessary for the successive optimisation of protein output. Thus, it is illustrated how a pronounced downside is overcome rather easily.

DISCUSSION

The absolute yield of protein is comparatively low in *C. acetobutylicum*, *E. coli* and *S. oneidensis*. This reflects the physiologic effort, biosynthesis of [FeFe] hydrogenases means to the host cell. Due to the most recent improvements for the isolation protocol of hydrogenase from *C. acetobutylicum* host cell culture, maximum yield is increased to 2.000 µg L^{-1} [139, 140]. However, normalised to $oD_{600nm} = 1$ the protein yield of *C. acetobutylicum* is still lower than with *E. coli* and *S. oneidensis*. Expression is under control of strong promoters and the recombinant hydrogenase is isolated via strep-tag affinity chromatography in each case. What makes a difference is that the clostridial host organism co-synthesises an intrinsic [FeFe] hydrogenase (*Ca*HydA) in addition to the recombinant one [35]. It is reasonable to assume that this dual burden restricts the overall yield of *Cr*HydA1 in *C. acetobutylicum*.

The *S. oneidensis* and *C. acetobutylicum* systems produce [FeFe] hydrogenases with high specific H_2 evolution activity. Both bacteria encode for intrinsic [FeFe] hydrogenases and have been shown to posses the specific maturation apparatus HydE/ F/ G. While *C. acetobutylicum* synthesises both *Ca*HydA and *Cr*HydA1, in case of *S. oneidensis*, a hydrogenase-deficient mutant is exploited. The double mutant *S. oneidensis* AS52 [137] is devoid of the [FeFe] and [NiFe] hydrogenase genes and completely lacks hydrogen activity. The transcription and protein level of HydE/ F/ G is not affected. The capability to produce [FeFe] hydrogenase with high specific activity is the most important argument in favour of the *S. oneidensis* and *C. acetobutylicum* systems.

Escherichia coli uses [NiFe] hydrogenases in mixed acid fermentation. No [FeFe] hydrogenase is found with *E. coli* [136]. To qualify the organism for maturation of [FeFe] hydrogenases, co-expression of HydE/F/G had to be achieved. At first, cloning the genes of *C. reinhardtii* HydEF, HydG, and *Cr*HydA1 resulted in negligible yield of hydrogenase protein (the maturases HydE and F are coupled to one protein HydEF in *C. reinhardtii*) [69]. King et al. eventually established the stabile co-expression of *C. acetobutylicum* HydE/ F/ G and *Cr*HydA1 in *E. coli* [134]. The absolute yield was found twice that of *S. oneidensis* but only minor values for V_{max} have been reported. Apparently, the cellular interplay of *Cr*HydA1 apoprotein and HydE/ F/ G gives active enzyme in only 16% of all cases. Although *E. coli* can not compete with *S. oneidensis* and *C. acetobutylicum* in terms of efficiency, the possibility to probe maturation with different combinations of HydE/ F/ G is an important and unique feature. The individual maturation enzymes and *in vivo* biogenesis of the H-cluster is best studied with *E. coli* [71, 72].

In summary, *S. oneidensis* AS52 hold the biggest potential in [FeFe] hydrogenase biosynthesis. The organism is genetically versatile and easily modified. Cell growth is performed in sealed vessels with no elaborate fermentation reactor necessary. *Shewanella oneidensis* AS52 transformants can be screened in a cell-based *in vitro* essay – the gram-negative organism readily disrupts upon osmotic shock. Mutants with maximal H_2 evolving activity can be chose for the production of recombinant

DISCUSSION

[FeFe] hydrogenases [139]. However, the successful isolation of a recombinant [FeFe] hydrogenase in *Shewanella* has been shown only once and the system needs essential optimisation before it can compete with the established *C. acetocutylicum* system. By now, *C. acetobutylicum* is the host cell system with the biggest outcome. Thanks to the improvements conducted in the last three years, an in-depth characterisation of *Ca*HydA and, above all, *Cr*HydA1 was made possible.

3.2 On the electronic structure of the H-cluster

Section (**2.3**) presents a closer look at the electronic structure of the H-cluster by X-ray absorption spectroscopy (XAS) at the k-edge of iron, more specifically *extended X-ray absorption fine structure* (EXAFS). The [FeFe] hydrogenase of *C. reinhardtii* facilitates an XAS analysis of the active site prosthetic group due to the absence of any ferredoxin-type [FeS] clusters. The overall structure of the *Cr*HydA1 H-cluster resembles the cofactor modelled into the crystal structures of *Cp*I and *Dd*H [32, 33, 55]. Sub-angstrom changes upon oxidative inhibition (CO) and reduction (H_2) were probed. Furthermore, we followed the effects of O_2 inactivation. The EXAFS characterisation of *Cr*HydA1 produced intricate structural data of high interest for synthetic chemistry and DFT modelling of the H-cluster [141-144].

It has been suggested that one of the terminal CO ligands at the distal iron atom of $[2Fe]_H$ shifts in a bridging position Fe_p–Fe_d upon oxidation of H_{red} to H_{ox} [32, 38]. After adding external CO to the reduced sample, the Fe_p–Fe_d distance increased about 0.1 Å supporting the formation of a "stiff" CO bridge in H_{ox}-CO (see Figure 10 for illustration). External CO is supposed to block the Fe_d binding site as discussed by various authors [31, 32, 38, 145] but not binding the Fe_p–Fe_d bridging position directly. The Fe–Fe distances in the [4Fe-4S] cluster are not affected by CO inhibition at the di-iron site.

Oxidation of the H-cluster with O_2 resulted in a quite surprising finding. Apparently, the contribution of the cubane cluster (Fe–Fe ≈ 2.7 Å) decreases faster upon inactivation than the absorption recorded for $[2Fe]_H$ (Fe–Fe ≈ 2.5/ 2.6 Å). The discrimination of this 0.1 Å difference was achieved by XAS examination for the first time and allowed following O_2 oxidation of both parts of the H-cluster separately. A discussion of this behaviour can be found in chapter (**3.4**).

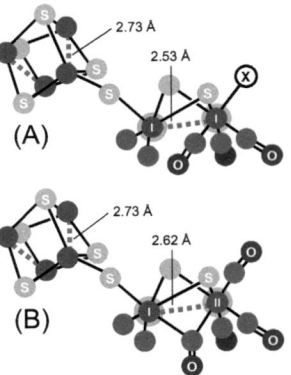

Figure 10 – Changes upon CO oxidation of the H-cluster as monitored by Fe–Fe distances (dashed lines). (A) H_{red} in the $Fe_p(I)$-$Fe_d(I)$ notation and (B) H_{ox}-CO with external CO bound to the vacant Fe_d site (X). While the distances of the iron atoms in the $[4Fe]_H$ cluster remain constant throughout the samples, CO binding causes a ~0.1 Å elongation of Fe–Fe in $[2Fe]_H$.

DISCUSSION

There is some uncertainty regarding the precise conformation of H_{red}. Successive gassing of dithionite-reduced CrHydA1 with H_2 did not result in distinct changes of the Fe–Fe distances. Binding of substrate H_2 in a Fe_p–Fe_d bridging position has been proposed [48, 146]. Two observations would have been indicative for this: (a) elongation of the Fe_p–Fe_d distance as in H_{ox}-CO and (b) no difference between H_{red} and H_{ox}-CO regarding the Fe_p–Fe_d [49, 146] distance. The former was not observed experimentally. On the contrary, the difference between H_{red} and H_{ox}-CO was remarkable (~ 0.1 Å). Thus, cooperative H_2 bonding to Fe_p–Fe_d bridging position can be ruled out.

Silakov et al. published a FTIR-voltammetric study of CrHydA using a "Moss" cell [147]. Upon reduction, the authors surprisingly found the bridging CO in H_{ox} still present in H_{red}. This notion has important impact for the discussion of the reaction mechanism as both oxidised and reduced states are catalytically active [29]. An intact CO bridge in H_{red} argues against the bridging position as hotspot of catalysis. This is in line with the observation of external CO being competitive with H_2 and O_2 [42] but contrasts with the structure of H_{red} as deducted from XAS. However, *in vitro* state-pure H_{red} is difficult to achieve. It is likely to assume that protein samples as derived from chemical and electro-chemical reduction are not the exactly same state. Note that the "super-reduced" state [45, 147, 148] has not been observed by protein film electrochemistry with CrHydA1.

3.3 Immobilisation of [FeFe] hydrogenases on conductive surfaces

Protein film electrochemistry (PFE) is used to study the activity of redox proteins in contact with a conductive surface. The [FeFe] hydrogenases DdH, CrHydA1, and CaHydA have been subject to elaborate protein film electrochemistry analyses. This is a representative examination of all variants of [FeFe] hydrogenases: *Desulvovibrio*, *Chlorophyta* and *Clostridia*-type enzymes are probed [31, 32, 60, 149]. Section **(2.2)** reports on immobilisation of CrHydA1 on a modified gold electrode and initial spectro-electrochemical investigations. Production of H_2 was probed by gas chromatography. Exploiting another system, protein film electrochemistry on *pyrolytic graphite edge* (PGE) is presented in **(2.4)** and **(2.5)**. Figure 11 shows a comparison of the different approaches.

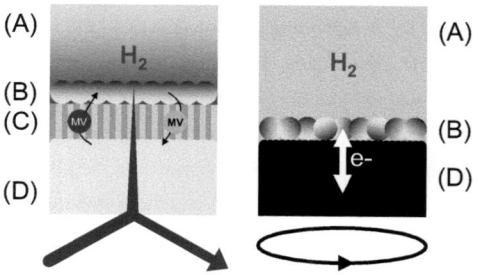

Figure 11 *– Schematic comparison of the set-up in SEIRAS (left) and graphite-based electrochemistry (right). (A) Working buffer, (B) protein film, (C) SAM, and (D) electrode. The shuttling methylviologen (MV) is drawn reduced (dark) and oxidised (light). On graphite, direct electron transfer takes place. Readings on graphite are performed with a RDE: While products (e.g., H_2) accumulate on stationary electrodes, rotation guarantees for diffusion-adjusted kinetics. Note the orientation in the protein film in the SEIRAS system.*

DISCUSSION

Spectro-electrochemical analysis of hydrogenase films

Immobilisation on a gold surface brings forth the chance to survey a protein film both by electrochemistry and *Fourier-transform infrared spectroscopy* (FTIR). The method applied is referred to as *surface-enhanced IR absorption spectroscopy* (SEIRAS) [150]. It makes use of plasmon enhancement of the IR beam at rough gold surfaces and the specific change of the absorption patterns if protein is bound to the surface [151]. Thus, electrochemical activity can directly be linked to structural / molecular changes of the protein. Infrared spectroscopy is a powerful technique in hydrogenase research in particular. The CN^- and CO ligands of the cofactors in [NiFe] and [FeFe] hydrogenases show specific absorbance from 2100 to 1800 cm^{-1}. The [FeFe] hydrogenases DdH and CpI have been characterised by FTIR [43, 45, 152], as well as DvH and the enzyme of *M. elsdenii* [153, 154]. Overall, bacterial [FeFe] hydrogenases share similar IR characteristics. Applying an in-solution spectro-electrochemical setup, CrHydA1 has been shown to behave distinctively different in FTIR [147].

CrHydA1 was probed on a *self-assembled monolayer* (SAM) of mercaptopropionic and mercaptoundecanoic acid. No direct electron transfer was observed – the activity of CrHydA1 relies on the presence of methylviologen as electron shuttle. Modification of gold surfaces is sufficient to avoid protein degradation. Making good benefit of this need, gold surfaces can be designed at will by using a wide range of linkers [150, 155], e.g. to help binding in a certain orientation. *Surface plasmon resonance* (SPR) was used to determine the concentration of bound protein. With mercaptopropionic acid as bio-compatible modification, H_2 evolution efficiency corresponds to about 15% of supplied electrons. Formation of the protein layer was followed by SEIRAS. However, the specific infrared signals of the H-cluster CN^- and CO ligands [43, 152] were not resolved.

The SEIRAS approach holds important advantages for the electrochemical-controlled infrared analysis of hydrogenases. Just recently, we were able to bind CrHydA1 to a short-length mercapto linker with viologen redox activity (unpublished experiments). Although this is still different from direct electron transfer, the potential control of bound protein is decoupled from methylviologen diffusion and allows for kinetic experiments with an improved temporal resolution. As most hydrogenases are able to accept electrons from methylviologen [16] we expect the presented surface-modification to work with hydrogenases from many species.

Direct electrochemistry of hydrogenase films on graphite

Other than with noble metal surfaces, proteins immediately bind to bio-compatible graphite [156]. Direct electron transfer occurs if the redox-active centre of the protein is in tunnelling distance to the conductive surface. Given this, current recorded as function of applied potential is a direct measure of

DISCUSSION

protein (redox) activity [157]. Armstrong et al. established protein film electrochemistry for many [NiFe] and [FeFe] hydrogenases using pyrolytic graphite edge rotating disc electrodes (see [24, 25] for review). In this set-up, hydrogenases bind the rough pyrolytic graphite surface without preset orientation. Direct electron transfer has been observed in most cases and the exposition of the protein film on the electrode surface allows for manifold chemical treatments.

Alternative systems have been introduced. Hambourger et al. used pyrolytic graphite and carbon felt electrodes [158] to construct a simple device for the photobiological production of H_2 [159]. Their approach incorporates a porphyrine-sensitised TiO_2 anode in electric contact with a cathode immobilised with CaHydA. In 2005, Rüdiger et al. came up with an approach that allows directed protein bonding to pyrolytic graphite surfaces. They modified an electrode with aminophenyl groups that can bind the negatively charged part of *D. gigas* [NiFe] hydrogenase. This domain is responsible for electron transfer [160]. More stable films and enhanced redox activity were achieved according to this modification.

The immobilisation of *Dd*H, *Cr*HydA1, and CaHydA on blank pyrolytic graphite rotating disc electrodes is reported in sections (**2.4**) and, more detailed, (**2.5**). The bacterial *Dd*H and CaHydA enzymes readily respond to the applied potential with sharp proton reducing activity. CaHydA shows only minor leaning towards H_2 uptake. *Dd*H and *Cr*HydA1 display both reduction and oxidation activity. Down to -600 mV vs. NHE, indications for a second electron transfer step (H_{red} → H_{sred} as in [45, 147]) were absent. Kinetic parameters support the notion that [FeFe] hydrogenases are typically not inhibited by H_2 as a product [13]. Interestingly, an area of sigmoid-shaped infliction was found around the midpoint potential of *Cr*HydA1. This reflects the special character of *Cr*HydA1 – due to the absence of the accessory [FeS] clusters, a little over-potential is necessary to drive electron transfer around -340 mV vs. NHE. Furthermore, *Cr*HydA1 and *Dd*H inactivate at potentials more positive than 0 V vs. NHE. "Anaerobic inactivation" at positive potentials is far less pronounced for CaHydA. This might indicate a correlation from catalytic activity and the extent of anaerobic inactivation. The process is reversible to different extent and typical for the hydrogenase probed [13]. A state "H_{ox}**inact**" is defined [161, 162] to distinguish this poorly understood state from H_{inact} and H_{ox}air [29].

3.4 Mechanisms of O_2 inactivation in [FeFe] hydrogenases

Oxygen sensitivity of hydrogenases is not a surprising thing to find if the general liability of [FeS] proteins to oxidative damage is taken into account (**1.2**). Any low-spin iron is oxidised rapidly upon contact with strong ligands like O_2 or CO. In parts this is because why ferrous iron has become a limiting nutrient in Plant and bacterial growth – most Fe(II) has been oxidised to Fe(III) under the influence of constantly increasing O_2 levels [163]. Iron-sulphur proteins have to deal with the same problem, so do hydrogenases.

DISCUSSION

"The appearance of oxygen on earth led to two major problems: the production of potentially deleterious reactive oxygen species and a drastic decrease in iron availability. In addition, iron, in its reduced form, potentates O_2 toxicity by converting (...) the less reactive H_2O_2 to more reactive oxygen species." Touati, 2000 [164]

The question is raised how hydrogenases modulate O_2 sensitivity. **First**, a structural comparison between [NiFe] and [FeFe] hydrogenases stresses the existence of gas channel filters in hydrogenases [76]. Carbon monoxide and O_2 are used to distinguish diffusion and reaction kinetics. **Second**, the molecular mechanisms of O_2 inactivation in [FeFe] hydrogenases are discussed. Inactivation has never been assayed intricately due to the small rate constant of inactivation k_{inact} – in our hands, as fast as 1.8×10^{-3} s^{-1} mM^{-1} for DdH. On the other hand, oxidation of [FeS] clusters is an effect that not only affects the H-cluster but all iron in the protein.

[NiFe] and [FeFe] hydrogenases display different levels of O_2 sensitivity

For the design of protein variants less sensitive to O_2 it is crucial to learn about the structural and molecular mechanisms of O_2 inactivation. Some strategies to circumvent O_2 damage in [FeS] proteins were presented in the introduction. Regarding their O_2 sensitivity, [NiFe] hydrogenases display the greatest versatility among all hydrogenases. Isolated under aerobiosis, standard-type, *Desulvovibrio* [NiFe] hydrogenases are in-sensitive to O_2 damage. The states "Ni-A" and "Ni-B" have been defined by EPR [29]. Due to an oxygen species blocking the Ni-Fe binding site the enzyme is inactive in this state [165]. Subsequent reduction gives the active "Ni-C" state that binds a hydride ion – and is prone to irreversible oxidative inhibition [166]. An important difference between [NiFe] and [FeFe] hydrogenases is that O_2 *inhibits* the enzyme. [FeFe] hydrogenases are *inactivated* and irreversibly unfolded, as suggested by circular dichroism spectroscopy (unpublished experiments). Carbonmonoxid blocks [NiFe] hydrogenases by attacking the terminal nickel site. "Ni-CO" refers to Ni-C as inhibited by CO [167]. In [FeFe] hydrogenases, O_2 and CO bind the same site (see below).

Remarkably, a certain subclass of [NiFe] hydrogenases is capable of hydrogen turnover in the presence of O_2 [168]. The membrane bound and soluble [NiFe] hydrogenases of Knallgas bacteria *Ralstonia eutropha* H16 and *R. metallidurans* CD34 were proven to function in H_2 cycling at ambient O_2 levels [13, 169]. The term "O_2 tolerance" was introduced; however it is not clear how these enzymes achieve to cope with O_2. An "oxidase activity" of the Ni-Fe cofactor has been discussed to be crucial in O_2 tolerance [25, 75].

The [NiFe] hydrogenases of *Ralstonia* species have been subject to intensive engineering. No crystal structure could have been obtained but sequence alignments suggest that Knallgas hydrogenases differ from "standard" [NiFe] hydrogenases in the position of two bulky residues

DISCUSSION

narrowing a gas channel that points to the catalytic nickel atom of the cofactor [168, 170, 171]. Replacing corresponding residues in the standard [NiFe] hydrogenases did result in a delayed inactivation but not O_2 tolerance [74, 172, 173]. Ludwig and co-workers tried to re-sensitise the O_2-tolerant [NiFe] hydrogenases of *R. eutropha* and *metallidurans* only to find these enzymes two orders *more* sensitive against O_2 than the wild type enzyme [174]. The authors conclude,

> "Tolerance to O_2 is clearly a complex factor and is determined by a well adapted spatial and electronic structure of the active site rather than a simple restriction of diffusion of inhibitory gases such as O_2." Ludwig et al., 2009 [174]

While connecting cavities are necessary to supply hydrogenases with substrate, a direct relation "tunnel width / O_2 sensitivity" is obviously not feasible [76]. In [FeFe] hydrogenases, only one gas channel ("A") has been found [31, 32]. Another is supported by DFT simulations ("B") but was not observed in crystal structures available [175]. In *Clostridia*, *Desulvovibrio*, and *Chlorophyta*-type hydrogenases, gas channel A is a conserved structural element [15]. This channel points to the distal iron atom in *Dd*H and *Ca*HydA, allowing for selective entrance only. Figure 12 shows the vicinity of the H-cluster in *Cp*I; note the enclosed surroundings of the $[4Fe]_H$ moiety. The flexibility of protein channels is subject to quite some discussion in gas-processing proteins [176, 177]. It might be likely that in hydrogenases, cavities form and deflate as a function of provided substrate – whether this is H_2, O_2 or CO [26, 76]. These considerations are important for a discussion of the pronounced variability in O_2 sensitivity we found with [FeFe] hydrogenases *Dd*H, *Cr*HydA1, and *Ca*HydA. The functional differences are not immediately evident from the (crystal) structures.

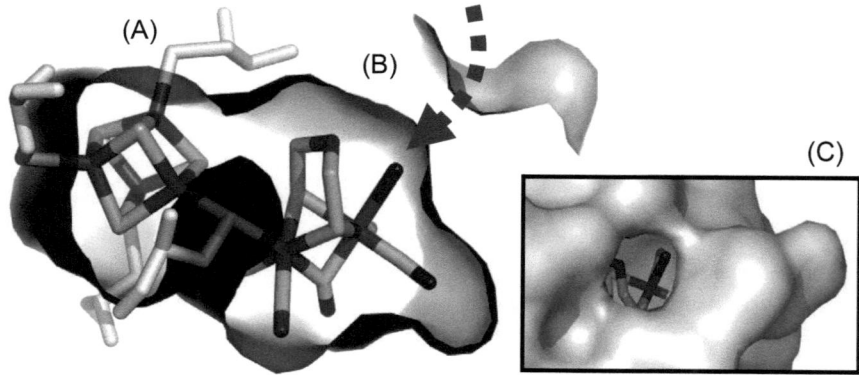

Figure 12 – *Selective access to the active site in [FeFe] hydrogenase CpI. (A) $[4Fe]_H$ subcluster with its coordinating cysteine residues (white). (B) $[2Fe]_H$ site facing the end of the gas channel "A" as found in the crystal structure. The azadithiolate ligand and the catalytic part of $[2Fe]_H$ (Fe_d, here blocked by CO) is oriented to the point where the cavity connects to the channel. (C) A look into the cavity from outside the immediate peptide environment of the H-cluster. Fe_d-CO and the azadithiolate nitrogen atom are visible.*

DISCUSSION

Oxygen sensitivity of three [FeFe] hydrogenases is probed by CO inhibition (**2.5**). Certain trends have been observed. Throughout all experiments, the notoriety for gas induced oxidation (expressed as k_{inact}) follows **DdH > CrHydA1 >> CaHydA**. The statement of Baffert et al. that "activity does not correlate with O_2 sensitivity" [133] was found being correct. The clostridial [FeFe] hydrogenase is less sensitive against oxidation by three orders of magnitude compared to DdH. Only minor discrimination between O_2 and CO was found. Note that CO inhibition is about 200-fold faster than O_2 inactivation in all three enzymes. The trend outlined above is mirrored in the differences between O_2 and CO when the catalytic direction is assayed. Oxygen inactivation is the same for CaHydA in H_2 oxidation and H^+ reduction direction (k_{inact} (O_2/H_2) / k_{inact} (O_2/H^+) ≈ 1) but DdH is 100 times more rapidly inactivated in H_2 oxidation. The same correlation was observed in CO inhibition. However, CO binds the H-cluster with greater affinity in the oxidised state. A preference of H_{ox} ($Fe_p(I)$ – $Fe_d(II)$) in CO bonding was argued before [25, 145]. The minimum ratio was k_{inact} (CO/H_2) / k_{inact} (CO/H^+) ≈ 5 for CaHydA.

It is likely to assume that the rate of catalysis and degree of susceptibility is *not* governed by gas channels. Despite the significant differences in inactivation and inhibition of the [FeFe] hydrogenases, O_2 / CO gas discrimination ratio is the same. Thus, not diffusion through the protein (k_{in} / k_{out} according to the kinetic model in (**2.5**)) but the actual binding process (k_2 / k_{-2}) determines the reaction rate. Similar, the rate of re-activation k_{re-act} from CO inhibition upon illumination [38, 148, 162] does not differ in DdH, CrHydA1, and CaHydA. Light scission of Fe–CO is not influenced by the active site environment [178] and k_{re-act} exclusively represents the rate of CO diffusing from the vicinity of the active site, and out of the protein. In Figure 13, k_{inact} for CO and O_2 are compared to k_{re-act} after illumination of the CO-inhibited electrode (adapted from Figure 9 in section (**2.5**)). Summed up, the different rates of inactivation in DdH, CrHydA1, and CaHydA do not stem from variably sized gas channels but reflect intrinsic parameters of the active site environment and the H-cluster.

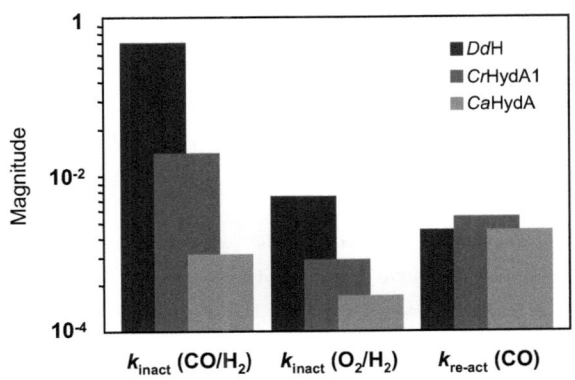

Figure 13 *– Comparison of k_{inact} for CO and O_2 and light-induced rate of reactivation after CO inhibition (k_{re-act}). Inactivation and inhibition was measured at H_2 uptake potential to avoid oxidation of O_2 on the graphite surface. While DdH, CrHydA1, and CaHyd differ by up to x1000 in k_{inact}, k_{re-act} is roughly the same for all hydrogenases. Note that inhibition (CO) is approximately 200-fold faster than inactivation (O_2).*

DISCUSSION

A model for the molecular mechanism of O_2 inactivation

Not much is known about the exact mechanisms of O_2 inactivation. Given that the extended protein scaffold of the H-cluster is not determining for the rate of inactivation, residues in the close vicinity of the cofactor are likely to determine inactivation, inhibition, and the rate of catalysis as well [85, 179, 180]. To elucidate the part of certain amino acids in O_2 inactivation, it is important to get an idea what exactly happens to O_2 when it approaches the H-cluster. We used a concerted XAS / electrochemistry analysis to reveal the mechanisms.

The electronic structure of the H-cluster in CrHydA1 was probed by XAS (**2.4**). The average distance of the iron atoms in $[4Fe]_H$ is 2.7 Å but slightly shorter in $[2Fe]_H$ (2.5 Å). The Fourier-transform EXAFS spectrum now reveals the selectivity of O_2 damage. Upon O_2 inactivation, the contribution of the Fe–Fe distance in $[4Fe]_H$ is reduced. The peak that represents Fe–Fe in $[2Fe]_H$ is remarkably stable. Inhibition with CO did not result in a decrease of these bands; same is true for H_2 reduction.

The different effects of O_2, CO and H_2 were assayed by protein film electrochemistry on pyrolytic graphite electrodes (**2.5**). Key to our observations was that (a) O_2 inactivation is prevented by CO inhibition and (b) the rate of inactivation is decreased in the presence of H_2. Thus, O_2, CO and H_2 competitively bind the same site [26, 28]. It is well established that external CO attacks the distal iron atom of the $[2Fe]_H$ site [38, 43, 145]. Accordingly, the vacant coordination site of this atom is where O_2 and H_2 do bind as well. Carbonmonoxide exclusively binds the $[2Fe]_H$ moiety. Oxidation of [4Fe4S] cluster by CO is chemically unlikely and has never been reported under the "mild" potentials applied here [181]. Given a near 100% protection of CO against O_2, we assume that O_2 selectively attacks the $[2Fe]_H$ site and not the cubane cluster. This is not trivial as [4Fe-4S] clusters are easily disrupted upon O_2 oxidation (**1.2**). Strong support for the impassivity of the $[4Fe]_H$ cluster comes from experiments on [FeFe] maturation.

The assembly of the H-cluster relies on a specific protein apparatus. Three enzymes have been identified that catalyze synthesis and translocation of the H-cluster: HydE, F, and G (**3.1**). According to an accepted proposal [182], HydF acts as a scaffold protein that holds a premature H-cluster formed by the help of radical-SAM proteins HydE and HydG. The latter is discussed to catalyse the synthesis of the azadithiolate ligand [183].

Just recently, HydF of *C. acetobutylicum* has been shown to contain a [FeS] cluster that displays characteristic CN^- and CO bands in the IR spectrum. This supports the theory of a premature H-cluster in HydF and the enzyme as a "construction site" of cofactor synthesis [73]. Mulder et al. could demonstrate that [FeFe] hydrogenase synthesised in *E. coli* (which naturally lacks the H-cluster maturation apparatus) contains not the H-cluster but a catalytically inactive [4Fe-4S] cluster [71]. This

DISCUSSION

apoprotein was shown to give nearly full hydrogen cycling activity after *in vitro* maturation with *Ca*HydF [73]. Mulder et al. reported the maturation of *Cr*HydA1 apoprotein as isolated under aerobiosis [71]. Obviously, the [4Fe-4S] cluster in the apoprotein is *not* perturbed upon O_2 contact. Maturation under anaerobiosis recovers not only activity but O_2 sensitivity as well. Active *Cr*HydA1 enzyme as inactivated by O_2 is impossible to re-activate by HydF or iron-sulphur cluster reconstitution [184] (unpublished experiments). This indicates damage of the $[2Fe]_H$ site *and* the cubane cluster.

Oxygen inactivation in [FeFe] hydrogenases can be explained with a three-step process. **First**, O_2 attacks the distal iron atom of the catalytic part of the H-cluster. **Second** is reduction of O_2. Fraser and co-workers discussed the "oxidase activity" of *Ralstonia*-type [NiFe] hydrogenases under O_2-rich conditions, pointing out that a similar process might be at hand in [FeFe] hydrogenases [25]. The difference, however, would be how the proteins deal with the product of O_2 reduction – which is, reactive oxygen species like peroxide or superoxide. Production of oxygen species by [FeS] clusters is a frequently observed phenomenon [83, 185]. **Third** could be (a) migration of reactive oxygen species to the nearby $[4Fe]_H$ cluster and subsequent degradation [114] or (b) "through-bond" oxidation $[4Fe]_H^{2+}$ → $[4Fe]_H^{3+}$ via the bridging cysteine. The $[4Fe]_H^{3+}$ state is typically not stable [186, 187] and loses an iron atom to form a transient $[3Fe-4S]^{+1}$ cluster. Both mechanisms initiate a stepwise degradation of $[4Fe]_H$ and, subsequently, the $[2Fe]_H$ cluster. A release of ferric iron (Fe(III)) to the medium is evident from the absorption edge in XAS for samples intensively treated with O_2. An example of controlled [4Fe-4S] cluster degradation via $[3Fe-4S]^{+1}$ can be found with FNR (**1.2**). Figure 14 illustrates the course of events in superoxide-mediated inactivation of [FeFe] hydrogenases.

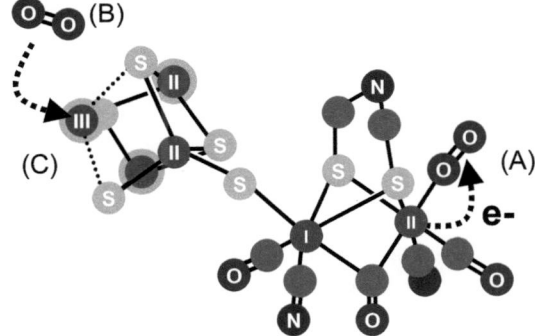

Figure 14 – Stepwise degradation of the H-cluster by superoxide. (A) Reduction of O_2 at the $[2Fe]_H$ site generates superoxide. (B) Superoxide (red) migrates and (C) oxidises the $[4Fe]_H$ cluster to an unstable +3 form. One iron atom (Fe(III)) is lost, and the $[3Fe-4S]^{+1}$ cluster dissolves spontaneously (not shown). Attack of the cubane subcluster is the first step in complete loss of the H-cluster. O_2 does not directly attack the $[4Fe]_H$ site.

If reactive oxygen species are discussed to be liable in O_2 inactivation, a comparison of [NiFe] and [FeFe] hydrogenases turns out to be helpful. Release of superoxide from the soluble [NiFe] hydrogenase of *R. eutropha* ("SH") upon aerobisation has been reported by Schneider and Schlegel 30 years ago [188]. Superoxide dismutase stabilises SH, catalase does not.

DISCUSSION

"The correlation between O_2 concentration (...) and inactivation rates and the stabilization of hydrogenase by addition of superoxide dismutase indicated that superoxide radicals are responsible for enzyme inactivation." Schneider and Schlegel, 1981 [188]

In contrast, the [FeFe] hydrogenase of *D. desulfuricans* is not protected against O_2 inactivation by superoxide dismutase or catalase [189]. Van der Westen et al. conclude that [FeFe] hydrogenases do not produce reactive oxygen species when gassed with O_2. We probed *Cr*HydA1 and SH for superoxide production and found the same trend verified (unpublished experiments). However, taking in account the molecular mechanism sketched above, superoxide "release" is not likely in [FeFe] hydrogenases. Due to the close proximity of $[2Fe]_H$ and $[4Fe]_H$ disruption of clusters should be rapid and inevitable [110].

We assume that [FeFe] hydrogenase can not deal with reactive oxygen species formed at the active site. No mechanism for reactive oxygen emission could evolve. From an evolutionary viewpoint, this was not necessary at all as [FeFe] hydrogenases replace O_2 as terminal electron acceptor in anaerobic respiration [190, 191]. The leaning towards H_2 evolution ("electron valve") reflects this function. [FeFe] hydrogenases have been found with strict anaerobes nearly exclusively [15, 16]. In green algae, they are synthesised only when O_2 is absent [59]. [FeFe] hydrogenases do not have to deal with O_2 – which is different in *Ralstonia*-type [NiFe] hydrogenases. These enzymes do not serve as "electron valves" but catalyse H_2 uptake under conditions of atmospheric O_2 [168]. Just recently, Saggu et al. published a spectroscopic study on the membrane bound [NiFe] hydrogenase of *R. eutropha* [192]. The authors found that,

"The proposed high potential species is therefore consistent with the presence of an additional iron close to the position of the proximal [4Fe4S] cluster or, alternatively, with the presence of an oxidized high potential [4Fe4S]$^{+3}$ at the proximal position." Saggu et al., 2009 [192]

A stable, high potential [4Fe4S] cluster would provide an environment insensitive against reactive oxygen damage [97]. This has prominently been shown for HiPIP ferredoxins (**1.2**). Thus, *Ralstonia*-type [NiFe] might have found a way to make use of a one-electron transition metal cofactor in the presence of O_2. Albeit the enzyme displays oxidase activity, reactive oxygen is guided "safely" out of the protein. The gas channel differences in *Ralstonia* and standard-type [NiFe] hydrogenase [75] are likely to modulate O_2 tolerance. However, key to catalysis under O_2 should be the cofactor environment rather than gas accessibility.

DISCUSSION

Towards the O_2-tolerant [FeFe] hydrogenase

The biotechnological generation of an O_2-tolerant [FeFe] hydrogenase is of greatest interest in basic and applied research. Two major findings should facilitate the design of novel hydrogenase variants: (a) O_2 sensitivity is not governed by gas channels and (b) the capability of the hydrogenases to deal with reactive oxygen species as product of O_2 reduction at the active site determines the level of O_2 tolerance.

To understand the molecular background of O_2 sensitivity in [FeFe] hydrogenases, a comparison of the active site environment in *Ca*HydA and *Dd*H will be helpful. The difference in O_2 sensitivity is 10^3 (**2.5**). Given that O_2 and CO reach the H-cluster in both enzymes at comparable rate, the underlying mechanism is not likely that of a gas filter. Three protein systems are considered to learn about O_2 sensitivity: High-potential ferredoxins (HiPIP), the soluble [NiFe] hydrogenase of *R. eutropha* (SH), and fumarate nitrate reduction (FNR). Scaffold-mediated stabilisation of the [4Fe-4S]$^{2+}$ → [4Fe-4S]$^{3+}$ transition in HiPIPs [97] could serve as an ideal for the residue composition around the [4Fe]$_H$ in [FeFe] hydrogenases. A high-potential subcluster would provide a tough environment for the release of reactive oxygen at the di-iron site – this is what has been discussed for the "Knallgas" hydrogenase SH [192]. Eventually, the course of event in cluster dismissal has to be taken in account again. The model depicted in (**2.4**) is quite similar to what has been reported for oxidative cluster degradation in FNR [132]. On the structural level, it is worthwhile to investigate how FNR is attacked by O_2 and subsequently deals with reactive oxygen species.

Due to the minimal [FeS] cluster composition, the [FeFe] hydrogenase of *C. reinhardtii* is the perfect subject for modifications towards an O_2 tolerant [FeFe] hydrogenase. Here, the heterologous synthesis in *S. oneidensis* AS52 holds potential for an efficient screening system. Other than with *C. acetobutylicum*, AS52 lacks any hydrogenase background so that a simple whole-cell *in vitro* test will yield immediate information on the catalytic profile of the recombinant hydrogenase.

SUMMARY AND ACKNOWLEDGEMENT

SUMMARY

This work elaborates on the molecular background of oxygen sensitivity in [FeFe] hydrogenases. Hydrogenases are iron-sulphur proteins that catalyse hydrogen turnover in micro organisms. [FeFe] hydrogenases in particular are high-efficiency catalysts for hydrogen evolution but irreversibly inactivated by O_2. Their unique [4Fe-4S]–[2Fe-2S] cofactor is referred to as "H-cluster".

Three different [FeFe] hydrogenases were analysed by electrochemistry. These were the bacterial hydrogenases *Ca*HydA and *Dd*H as well as *Cr*HydA1 of *Chlamydomonas reinhardtii*, a unicellular green alga. The effects of O_2 and CO were probed, the latter of which reversibly inhibits hydrogenases. *Dd*H is inactivated by O_2 ten times more rapid than *Cr*HydA1 (2.2×10^{-4} s^{-1} µM^{-1}) and three orders of magnitude faster than *Ca*HydA (5.5×10^{-6} s^{-1} µM^{-1}). The same correlation was found for CO inhibition, although reactivation is in the same range for all three enzymes. Moreover, CO was shown to prevent the H-cluster from O_2 damage. Given that CO can not bind the [4Fe-4S] part of the cofactor, the capability of CO to protect the entire H-cluster proves that O_2 attacks the [2Fe-2S] site exclusively, too.

The main focus of this study was the structural analysis of the H-cluster of *Cr*HydA1. For the first time, the H-cluster was analysed by X-ray absorption spectroscopy (XAS) at the K-edge of iron. Due to the accuracy of XAS, resolution even of closely spaced interatomic distances was achieved. The *Cr*HydA1 H-cluster consists of a [4Fe4S] cluster (Fe–Fe ≈ 2.53 Å) and a di-iron site (Fe–Fe ≈ 2.73 Å) which both are similar to their crystallographically characterized counterparts. Treatment of *Cr*HydA1 with CO resulted in a 0.1 Å increase of the Fe–Fe distance of the [2Fe-2S] site. This agrees well with formation of a bridging CO in H_{ox}-CO. Surprisingly, XAS shows that reaction with O_2 results in destruction of the [4Fe-4S] domain of the active site H-cluster while leaving the di-iron moiety essentially intact.

The XAS analysis of *Cr*HydA1 represents the first experimental approach to the structural implications of O_2 inactivation. Although electrochemistry has shown that O_2 does not attack the [4Fe-4S] part of the active site cofactor, XAS data suggests specific degradation of the cubane cluster. Illustrating the course of events, a novel thesis on O_2 inactivation of [FeFe] hydrogenases is devised.

SUMMARY AND ACKNOWLEDGEMENT

ACKNOWLEDGEMENT

First off, a few words in English. I would like to thank the numerous people I was allowed to work in cooperation with. **PD Dr. Michael Haumann**, for the nightshifts at the Synchrotron in Hamburg and Berlin, intense discussions and late night lunches. **Prof. Dr. Fraser A. Armstrong**, for letting me mess around with the Braun Box and answering endless emails of speculation and guesswork. **Dr. Henning Krassen**, for gigabytes of lost FTIR data, taking care of the hydrofluoric acid and Mario Kart. Of course, **Dr. Gabrielle Goldet**: for the months she spent surveying touchy protein films and all the proofreading I asked her to do.

Natürlich gebührt der erste Dank meinem Doktorvater **Prof. Dr. Thomas Happe**, der mich seit meiner Diplomzeit immer unterstützt hat und dessen Forschungsprojekte ich mitgestalten durfte. Danken möchte ich zudem meinem Korreferenten **Prof. Dr. Eckhard Hofmann**, außerdem **Prof. Dr. Achim Trebst** für Kritik, Diskussion und Ratschlag.

Besonders versüßt hat mir die Promotionszeit die andauernde Freundschaft zu **Gregory** und **Thilo**. Ohne Euch wäre es nur halb so schön gewesen! Unverzichtbar der Ratschlag von **Martin**, der *gossip* von **Camilla** und **Jessica**, sowie die Unterhaltungen zur Nacht mit **Gabriele**. Viel Glück auf Euren Wegen meinen Diplomanten **Sebastian**, **Lukas** und ~~Jennis~~ **Dennis**. Überhaupt das ganze „Happy Lab" – Ich fürchte, eine so nette Arbeitsgruppe findet sich nicht wieder! Der Platz reicht nicht für die Nennung aller Namen. Ein Extra Dank an **Adrian** – wer weiß, ob ich ohne Dich an die Elektrochemie gekommen wäre?

Danke **Mama**, **Papa**, **Svenja** und **Paul** für Alles. Nicht weniger. Was soll ich sonst sagen? Erwähnt sollen auch **Monika**, **Werner** und **Lisa** sein, die mir nicht nur ihre Tochter und Schwester anvertraut, sondern mich auch als Sohn und Bruder ins Herz geschlossen haben.

Diese Arbeit ist meiner eigenen kleinen Familie gewidmet – meinem Söhnchen **Ari** und seiner schönen, schlauen, liebevollen Mutter **Anne**. *You rock my world*, noch immer.

LITERATURE

1. Sørensen, B.M., *Hydrogen Conversion*, in *Hydrogen And Fuel Cells: Emerging Technologies And Applications (sustainable World)*, B.M. Sørensen and B. Srensen, Editors. 2005, Elsevier Academic Press. p. 76-82.
2. Kasting, J.F., *Earth's early atmosphere.* Science, 1993. **259**(5097): p. 920-6.
3. Tian, F., et al., *A hydrogen-rich early Earth atmosphere.* Science, 2005. **308**(5724): p. 1014-7.
4. Kasting, J.F., *Earth history. The rise of atmospheric oxygen.* Science, 2001. **293**(5531): p. 819-20.
5. Nisbet, E.G. and N.H. Sleep, *The habitat and nature of early life.* Nature, 2001. **409**(6823): p. 1083-91.
6. Wu, M., et al., *Life in hot carbon monoxide: the complete genome sequence of Carboxydothermus hydrogenoformans Z-2901.* PLoS Genet, 2005. **1**(5): p. e65.
7. Svetlitchnyi, V., et al., *Two membrane-associated NiFeS-carbon monoxide dehydrogenases from the anaerobic carbon-monoxide-utilizing eubacterium Carboxydothermus hydrogenoformans.* J Bacteriol, 2001. **183**(17): p. 5134-44.
8. Leigh, J.A., *Nitrogen fixation in methanogens: the archaeal perspective.* Curr Issues Mol Biol, 2000. **2**(4): p. 125-31.
9. Thiel, T. and B. Pratte, *Effect on heterocyst differentiation of nitrogen fixation in vegetative cells of the cyanobacterium Anabaena variabilis ATCC 29413.* J Bacteriol, 2001. **183**(1): p. 280-6.
10. Meeks, J.C., et al., *An overview of the genome of Nostoc punctiforme, a multicellular, symbiotic cyanobacterium.* Photosynth Res, 2001. **70**(1): p. 85-106.
11. Bowien, B. and H.G. Schlegel, *Physiology and biochemistry of aerobic hydrogen-oxidizing bacteria.* Annu Rev Microbiol, 1981. **35**: p. 405-52.
12. Brandis, A. and R.K. Thauer, *Growth of Desulfovibrio Species on Hydrogen and Sulfate as Sole Energy-Source.* Journal of General Microbiology, 1981. **126**(SEP): p. 249-252.
13. Vincent, K.A., et al., *Electrochemical definitions of O2 sensitivity and oxidative inactivation in hydrogenases.* J Am Chem Soc, 2005. **127**(51): p. 18179-89.
14. Friedrich, B. and E. Schwartz, *Molecular biology of hydrogen utilization in aerobic chemolithotrophs.* Annu Rev Microbiol, 1993. **47**: p. 351-83.
15. Vignais, P.M. and B. Billoud, *Occurrence, classification, and biological function of hydrogenases: an overview.* Chem Rev, 2007. **107**(10): p. 4206-72.
16. Adams, M.W.W., *The structure and mechanism of iron-hydrogenases.* Biochimica et Biophysica Acta, 1990. **1020**: p. 115-145.
17. Zirngibl, C., et al., *H_2-forming methylenetetrahydromethanopterin dehydrogenase, a novel type of hydrogenase without iron-sulfur clusters in methanogenic archaea.* Eur J Biochem, 1992. **208**(2): p. 511-20.
18. Shima, S., et al., *The crystal structure of [Fe]-hydrogenase reveals the geometry of the active site.* Science, 2008. **321**(5888): p. 572-5.
19. Hiromoto, T., et al., *The crystal structure of an [Fe]-hydrogenase-substrate complex reveals the framework for H2 activation.* Angew Chem Int Ed Engl, 2009. **48**(35): p. 6457-60.
20. Shima, S. and R.K. Thauer, *A third type of hydrogenase catalyzing H2 activation.* Chem Rec, 2007. **7**(1): p. 37-46.
21. Hiromoto, T., et al., *The crystal structure of C176A mutated [Fe]-hydrogenase suggests an acyl-iron ligation in the active site iron complex.* FEBS Lett, 2009. **583**(3): p. 585-90.
22. Schleucher, J., et al., *H2-forming N5, N10-methylenetetrahydromethanopterin dehydrogenase from Methanobacterium thermoautotrophicum catalyzes a stereoselective hydride transfer as determined by two-dimensional NMR spectroscopy.* Biochemistry, 1994. **33**(13): p. 3986-93.
23. Lyon, E.J., et al., *UV-A/blue-light inactivation of the 'metal-free' hydrogenase (Hmd) from methanogenic archaea.* Eur J Biochem, 2004. **271**(1): p. 195-204.
24. Vincent, K.A., A. Parkin, and F.A. Armstrong, *Investigating and exploiting the electrocatalytic properties of hydrogenases.* Chem Rev, 2007. **107**(10): p. 4366-413.

LITERATURE

25. Armstrong, F.A., et al., *Dynamic electrochemical investigations of hydrogen oxidation and production by enzymes and implications for future technology.* Chem Soc Rev, 2009. **38**(1): p. 36-51.
26. Fontecilla-Camps, J.C., et al., *Structure/function relationships of [NiFe]- and [FeFe]- hydrogenases.* Chem Rev, 2007. **107**(10): p. 4273-303.
27. Ogata, H., W. Lubitz, and Y. Higuchi, *[NiFe] hydrogenases: structural and spectroscopic studies of the reaction mechanism.* Dalton Trans, 2009(37): p. 7577-87.
28. De Lacey, A.L., et al., *Activation and inactivation of hydrogenase function and the catalytic cycle: spectroelectrochemical studies.* Chem Rev, 2007. **107**(10): p. 4304-30.
29. Lubitz, W., E. Reijerse, and M. van Gastel, *[NiFe] and [FeFe] hydrogenases studied by advanced magnetic resonance techniques.* Chem Rev, 2007. **107**(10): p. 4331-65.
30. Frey, M., *Hydrogenases: hydrogen-activating enzymes.* Chembiochem, 2002. **3**(2-3): p. 153-60.
31. Peters, J.W., et al., *X-ray crystal structure of the Fe-only hydrogenase (CpI) from Clostridium pasteurianum to 1.8 angstrom resolution.* Science, 1998. **282**(5395): p. 1853-8.
32. Nicolet, Y., et al., *Desulfovibrio desulfuricans iron hydrogenase: the structure shows unusual coordination to an active site Fe binuclear center.* Structure Fold Des, 1999. **7**(1): p. 13-23.
33. Pandey, A.S., et al., *Dithiomethylether as a ligand in the hydrogenase h-cluster.* J Am Chem Soc, 2008. **130**(13): p. 4533-40.
34. Girbal, L., et al., *Development of a sensitive gene expression reporter system and an inducible promoter-repressor system for Clostridium acetobutylicum.* Appl Environ Microbiol, 2003. **69**(8): p. 4985-8.
35. Girbal, L., et al., *Homologous and heterologous overexpression in Clostridium acetobutylicum and characterization of purified clostridial and algal Fe-only hydrogenases with high specific activities.* Appl Environ Microbiol, 2005. **71**(5): p. 2777-81.
36. Stokkermans, J.P., et al., *Overproduction of prismane protein in Desulfovibrio vulgaris (Hildenborough): evidence for a second S = 1/2-spin system in the one-electron reduced state.* Eur J Biochem, 1992. **210**(3): p. 983-8.
37. Pierik, A.J., et al., *Redox properties of the iron-sulfur clusters in activated Fe-hydrogenase from Desulfovibrio vulgaris (Hildenborough).* Eur J Biochem, 1992. **209**(1): p. 63-72.
38. Lemon, B.J. and J.W. Peters, *Binding of exogenously added carbon monoxide at the active site of the iron-only hydrogenase (CpI) from Clostridium pasteurianum.* Biochemistry, 1999. **38**(40): p. 12969-73.
39. Silakov, A., et al., *14N HYSCORE investigation of the H-cluster of [FeFe] hydrogenase: evidence for a nitrogen in the dithiol bridge.* Phys. Chem. Chem. Phys., 2009: p. in press.
40. Silakov, A., et al., *The electronic structure of the H-cluster in the [FeFe]-hydrogenase from Desulfovibrio desulfuricans: a Q-band 57Fe-ENDOR and HYSCORE study.* J Am Chem Soc, 2007. **129**(37): p. 11447-58.
41. Pierik, A.J., et al., *Multi-frequency EPR and high-resolution Mossbauer spectroscopy of a putative [6Fe-6S] prismane-cluster-containing protein from Desulfovibrio vulgaris (Hildenborough). Characterization of a supercluster and superspin model protein.* Eur J Biochem, 1992. **206**(3): p. 705-19.
42. Hatchikian, E.C., et al., *Further characterization of the [Fe]-hydrogenase from Desulfovibrio desulfuricans ATCC 7757.* Eur J Biochem, 1992. **209**(1): p. 357-65.
43. Nicolet, Y., et al., *Crystallographic and FTIR spectroscopic evidence of changes in Fe coordination upon reduction of the active site of the Fe-only hydrogenase from Desulfovibrio desulfuricans.* J Am Chem Soc, 2001. **123**(8): p. 1596-601.
44. Carepo, M., et al., *17O ENDOR detection of a solvent-derived Ni-(OH(x))-Fe bridge that is lost upon activation of the hydrogenase from Desulfovibrio gigas.* J Am Chem Soc, 2002. **124**(2): p. 281-6.
45. Roseboom, W., et al., *The active site of the [FeFe]-hydrogenase from Desulfovibrio desulfuricans. II. Redox properties, light sensitivity and CO-ligand exchange as observed by infrared spectroscopy.* J Biol Inorg Chem, 2006. **11**(1): p. 102-18.
46. Foerster, S., et al., *An orientation-selected ENDOR and HYSCORE study of the Ni-C active state of Desulfovibrio vulgaris Miyazaki F hydrogenase.* Journal of Biological Inorganic Chemistry, 2005. **10**(1): p. 51-62.

LITERATURE

47. Brecht, M., et al., *Direct detection of a hydrogen ligand in the [NiFe] center of the regulatory H-2-sensing hydrogenase from Ralstonia eutropha in its reduced state by HYSCORE and ENDOR spectroscopy.* Journal of the American Chemical Society, 2003. **125**(43): p. 13075-13083.
48. Bruschi, M., P. Fantucci, and L. De Gioia, *Density functional theory investigation of the active site of [Fe]-hydrogenases: effects of redox state and ligand characteristics on structural, electronic, and reactivity properties of complexes related to the [2Fe]H subcluster.* Inorg Chem, 2003. **42**(15): p. 4773-81.
49. Zhou, T., et al., *Density functional study on dihydrogen activation at the H cluster in Fe-only hydrogenases.* Inorg Chem, 2005. **44**(14): p. 4941-6.
50. Fan, H.J. and M.B. Hall, *A capable bridging ligand for Fe-only hydrogenase: density functional calculations of a low-energy route for heterolytic cleavage and formation of dihydrogen.* J Am Chem Soc, 2001. **123**(16): p. 3828-9.
51. Liu, Z.P. and P. Hu, *A density functional theory study on the active center of Fe-only hydrogenase: characterization and electronic structure of the redox states.* J Am Chem Soc, 2002. **124**(18): p. 5175-82.
52. Tye, J.W., et al., *Dual electron uptake by simultaneous iron and ligand reduction in an N-heterocyclic carbene substituted [FeFe] hydrogenase model compound.* Inorg Chem, 2005. **44**(16): p. 5550-2.
53. Darensbourg, M.Y., et al., *The organometallic active site of [Fe]hydrogenase: models and entatic states.* Proc Natl Acad Sci U S A, 2003. **100**(7): p. 3683-8.
54. Atta, M. and J. Meyer, *Characterization of the gene encoding the [Fe]-hydrogenase from Megasphaera elsdenii.* Biochim Biophys Acta, 2000. **1476**(2): p. 368-71.
55. Peters, J.W., *Structure and mechanism of iron-only hydrogenases.* Curr Opin Struct Biol, 1999. **9**(6): p. 670-6.
56. Kessler, E., *Effect of anaerobiosis on photosynthetic reactions and nitrogen metabolism of algae with and without hydrogenase.* Arch Mikrobiol, 1973. **93**(2): p. 91-100.
57. Winkler, M., et al., *[Fe]-hydrogenases in green algae: photo-fermentation and hydrogen evolution under sulfur deprivation.* International Journal of Hydrogen Energy, 2002. **27**: p. 1431-1439.
58. Schnackenberg, J., R. Schulz, and H. Senger, *Characterization and purification of a hydrogenase from the eukaryotic green alga Scenedesmus obliquus.* FEBS Lett, 1993. **327**(1): p. 21-4.
59. Happe, T. and A. Kaminski, *Differential regulation of the Fe-hydrogenase during anaerobic adaptation in the green alga Chlamydomonas reinhardtii.* Eur J Biochem, 2002. **269**(3): p. 1022-32.
60. Happe, T. and J.D. Naber, *Isolation, characterization and N-terminal amino acid sequence of hydrogenase from the green alga Chlamydomonas reinhardtii.* Eur J Biochem, 1993. **214**(2): p. 475-81.
61. Schmitter, J.M., et al., *Purification, properties and complete amino acid sequence of the ferredoxin from a green alga, Chlamydomonas reinhardtii.* Eur J Biochem, 1988. **172**(2): p. 405-12.
62. Melis, A., et al., *Sustained photobiological hydrogen gas production upon reversible inactivation of oxygen evolution in the green alga Chlamydomonas reinhardtii.* Plant Physiol, 2000. **122**(1): p. 127-36.
63. Ghirardi, M.L., et al., *Photobiological hydrogen-producing systems.* Chem Soc Rev, 2009. **38**(1): p. 52-61.
64. Ghirardi, M.L., et al., *Approaches to developing biological H(2)-photoproducing organisms and processes.* Biochem Soc Trans, 2005. **33**(Pt 1): p. 70-2.
65. Ghirardi, M.L., et al., *Hydrogenases and Hydrogen Photoproduction in Oxygenic Photosynthetic Organisms.* Annu Rev Plant Biol, 2006.
66. Ghirardi, M.L., et al., *Microalgae: a green source of renewable H(2).* Trends Biotechnol, 2000. **18**(12): p. 506-11.
67. Winkler, M., et al., *Characterization of the key step for light-driven hydrogen evolution in green algae.* J Biol Chem, 2009. **284**(52): p. 36620-7.

LITERATURE

68. Kamp, C., et al., *Isolation and first EPR characterization of the [FeFe]-hydrogenases from green algae.* Biochim Biophys Acta, 2008. **1777**(5): p. 410-416.
69. Posewitz, M.C., et al., *Discovery of two novel radical S-adenosylmethionine proteins required for the assembly of an active [Fe] hydrogenase.* J Biol Chem, 2004. **279**(24): p. 25711-20.
70. Sybirna, K., et al., *Shewanella oneidensis: a new and efficient system for expression and maturation of heterologous [Fe-Fe] hydrogenase from Chlamydomonas reinhardtii.* BMC Biotechnol, 2008. **8**: p. 73.
71. Mulder, D.W., et al., *Activation of HydA(DeltaEFG) requires a preformed [4Fe-4S] cluster.* Biochemistry, 2009. **48**(26): p. 6240-8.
72. McGlynn, S.E., et al., *In vitro activation of [FeFe] hydrogenase: new insights into hydrogenase maturation.* J Biol Inorg Chem, 2007. **12**(4): p. 443-7.
73. Czech, I., et al., *The [FeFe]-hydrogenase maturase HydF from Clostridium acetobutylicum contains a CO and CN(-) ligated iron cofactor.* FEBS Lett, 2009.
74. Leroux, F., et al., *Experimental approaches to kinetics of gas diffusion in hydrogenase.* Proc Natl Acad Sci U S A, 2008. **105**(32): p. 11188-93.
75. Cracknell, J.A., et al., *A kinetic and thermodynamic understanding of O2 tolerance in [NiFe]-hydrogenases.* Proc Natl Acad Sci U S A, 2009.
76. Fontecilla-Camps, J.C., et al., *Structure-function relationships of anaerobic gas-processing metalloenzymes.* Nature, 2009. **460**(7257): p. 814-22.
77. Edward, I.S., et al., *Ligand K-edge X-ray absorption spectroscopy: covalency of ligand–metal bonds.* Coordination Chemistry Reviews, 2004. **249**(1-2): p. 97-129.
78. Rees, D.C., *Great metalloclusters in enzymology.* Annu Rev Biochem, 2002. **71**: p. 221-46.
79. Capozzi, F., S. Ciurli, and C. Luchinat, *Coordination sphere versus protein environment as determinants of electronic and functional properties of iron-sulfur proteins.* Metal Sites in Proteins and Models, 1998. **90**: p. 127-160.
80. Babini, E., et al., *Experimental evidence for the role of buried polar groups in determining the reduction potential of metalloproteins: the S79P variant of Chromatium vinosum HiPIP.* Journal of Biological Inorganic Chemistry, 1999. **4**(6): p. 692-700.
81. Bendall, D.S., *The Unfinished Story of Cytochrome f.* Photosynth Res, 2004. **80**(1-3): p. 265-76.
82. Takahama, U., M. Shimizutakahama, and U. Heber, *The Redox State of the Nadp System in Illuminated Chloroplasts.* Biochimica Et Biophysica Acta, 1981. **637**(3): p. 530-539.
83. Imlay, J.A., *Iron-sulphur clusters and the problem with oxygen.* Mol Microbiol, 2006. **59**(4): p. 1073-82.
84. Dey, A., et al., *Ligand K-edge X-ray absorption spectroscopy and DFT calculations on [Fe3S4]0,+ clusters: delocalization, redox, and effect of the protein environment.* J Am Chem Soc, 2004. **126**(51): p. 16868-78.
85. Dey, A., et al., *Solvent tuning of electrochemical potentials in the active sites of HiPIP versus ferredoxin.* Science, 2007. **318**(5855): p. 1464-8.
86. Niu, S. and T. Ichiye, *Insight into environmental effects on bonding and redox properties of [4Fe-4S] clusters in proteins.* J Am Chem Soc, 2009. **131**(16): p. 5724-5.
87. Daniel, R.M. and M.J. Danson, *Did primitive microorganisms use nonhem iron proteins in place of NAD/P?* J. Mol. Evol., 1995. **40**: p. 559-63.
88. Huber, C. and G. Wachtershauser, *Peptides by activation of amino acids with CO on (Ni,Fe)S surfaces: Implications for the origin of life.* Science, 1998. **281**(5377): p. 670-672.
89. Hagen, K.S., J.G. Reynolds, and R.H. Holm, *Definition of Reaction Sequences Resulting in Self-Assembly of [Fe4s4(Sr)4]2- Clusters from Simple Reactants.* Journal of the American Chemical Society, 1981. **103**(14): p. 4054-4063.
90. Venkateswara Rao, P. and R.H. Holm, *Synthetic analogues of the active sites of iron-sulfur proteins.* Chem Rev, 2004. **104**(2): p. 527-59.
91. Matsubara, H., et al., *Structural and evolution of chloroplast- and bacterial-type ferredoxins.* UCLA Forum Med Sci, 1979(21): p. 245-66.
92. Rypniewski, W.R., et al., *Crystallization and Structure Determination to 2.5-a Resolution of the Oxidized [Fe2-S2] Ferredoxin Isolated from Anabaena-7120.* Biochemistry, 1991. **30**(17): p. 4126-4131.

LITERATURE

93. Bertini, I., S. Ciurli, and C. Luchinat, *The Electronic-Structure of Fes Centers in Proteins and Models a Contribution to the Understanding of Their Electron-Transfer Properties*. Iron-Sulfur Proteins Perovskites, 1995. **83**: p. 1-53.
94. Bertini, I., et al., *Solution Structure of the Oxidized 2[4fe-4s] Ferredoxin from Clostridium-Pasteurianum*. European Journal of Biochemistry, 1995. **232**(1): p. 192-205.
95. Rieske, J.S., D.H. Maclennan, and R. Coleman, *Isolation + Properties of Iron-Protein from (Reduced Coenzyme Q) -Cytochrome C Reductase Complex of Respiratory Chain*. Biochemical and Biophysical Research Communications, 1964. **15**(4): p. 338-&.
96. Denke, E., et al., *Alteration of the midpoint potential and catalytic activity of the Rieske iron-sulfur protein by changes of amino acids forming hydrogen bonds to the iron-sulfur cluster*. Journal of Biological Chemistry, 1998. **273**(15): p. 9085-9093.
97. Bertini, I., et al., *H-1 and C-13 NMR studies of an oxidized HiPIP*. Inorganic Chemistry, 1997. **36**(21): p. 4798-4803.
98. Agarwal, A., D. Li, and J.A. Cowan, *Role of aromatic residues in stabilization of the [Fe4S4] cluster in high-potential iron proteins (HiPIPs): physical characterization and stability studies of Tyr-19 mutants of Chromatium vinosum HiPIP*. Proc Natl Acad Sci U S A, 1995. **92**(21): p. 9440-4.
99. Glaser, T., et al., *Protein effects on the electronic structure of the [Fe4S4]2+ cluster in ferredoxin and HiPIP*. J Am Chem Soc, 2001. **123**(20): p. 4859-60.
100. Lauble, H., et al., *Crystal structures of aconitase with isocitrate and nitroisocitrate bound*. Biochemistry, 1992. **31**(10): p. 2735-48.
101. Beinert, H. and P.J. Kiley, *Fe-S proteins in sensing and regulatory functions*. Curr Opin Chem Biol, 1999. **3**(2): p. 152-7.
102. Johnson, D.C., et al., *Structure, function, and formation of biological iron-sulfur clusters*. Annu Rev Biochem, 2005. **74**: p. 247-81.
103. Sofia, H.J., et al., *Radical SAM, a novel protein superfamily linking unresolved steps in familiar biosynthetic pathways with radical mechanisms: functional characterization using new analysis and information visualization methods*. Nucleic Acids Res, 2001. **29**(5): p. 1097-106.
104. Nicolet, Y. and C.L. Drennan, *AdoMet radical proteins--from structure to evolution-- alignment of divergent protein sequences reveals strong secondary structure element conservation*. Nucleic Acids Res, 2004. **32**(13): p. 4015-25.
105. Frey, P.A., A.D. Hegeman, and F.J. Ruzicka, *The Radical SAM Superfamily*. Crit Rev Biochem Mol Biol, 2008. **43**(1): p. 63-88.
106. Chen, D., et al., *Coordination and mechanism of reversible cleavage of S-adenosylmethionine by the [4Fe-4S] center in lysine 2,3-aminomutase*. J Am Chem Soc, 2003. **125**(39): p. 11788-9.
107. Kulzer, R., et al., *Reconstitution and characterization of the polynuclear iron-sulfur cluster in pyruvate formate-lyase-activating enzyme. Molecular properties of the holoenzyme form*. J Biol Chem, 1998. **273**(9): p. 4897-903.
108. Ugulava, N.B., B.R. Gibney, and J.T. Jarrett, *Biotin synthase contains two distinct iron-sulfur cluster binding sites: chemical and spectroelectrochemical analysis of iron-sulfur cluster interconversions*. Biochemistry, 2001. **40**(28): p. 8343-51.
109. Naqui, A., B. Chance, and E. Cadenas, *Reactive oxygen intermediates in biochemistry*. Annu Rev Biochem, 1986. **55**: p. 137-66.
110. Halliwell, B., *Reactive species and antioxidants. Redox biology is a fundamental theme of aerobic life*. Plant Physiology, 2006. **141**(2): p. 312-322.
111. Broderick, J.B., et al., *Pyruvate formate-lyase activating enzyme is an iron-sulfur protein*. Journal of the American Chemical Society, 1997. **119**(31): p. 7396-7397.
112. Cosper, M.M., et al., *Characterization of the cofactor composition of Escherichia coli biotin synthase*. Biochemistry, 2004. **43**(7): p. 2007-2021.
113. Crack, J.C., et al., *Superoxide-mediated amplification of the oxygen-induced switch from [4Fe-4S] to [2Fe-2S] clusters in the transcriptional regulator FNR*. Proc Natl Acad Sci U S A, 2007. **104**(7): p. 2092-7.

LITERATURE

114. Tilley, G.J., et al., *Influence of electrochemical properties in determining the sensitivity of [4Fe-4S] clusters in proteins to oxidative damage.* Biochemical Journal, 2001. **360**: p. 717-726.
115. Koay, M.S., et al., *Modelling low-potential [Fe4S4] clusters in proteins.* Chemistry & Biodiversity, 2008. **5**(8): p. 1571-1587.
116. Lhee, S., et al., *Modifications of protein environment of the [2Fe-2S] cluster of the bc1 complex: Effects on the biophysical properties of the Rieske iron-sulfur protein and on the kinetics of the complex.* J Biol Chem, 2009.
117. Jordan, P., et al., *Three-dimensional structure of cyanobacterial photosystem I at 2.5 A resolution.* Nature, 2001. **411**(6840): p. 909-17.
118. Sazanov, L.A., *Respiratory complex I: Mechanistic and structural insights provided by the crystal structure of the hydrophilic domain.* Biochemistry, 2007. **46**(9): p. 2275-2288.
119. Lill, R., *Function and biogenesis of iron-sulphur proteins.* Nature, 2009. **460**(7257): p. 831-8.
120. Green, J., et al., *Bacterial sensors of oxygen.* Curr Opin Microbiol, 2009. **12**(2): p. 145-51.
121. Tard, C. and C.J. Pickett, *Structural and functional analogues of the active sites of the [Fe]-, [NiFe]-, and [FeFe]-hydrogenases.* Chem Rev, 2009. **109**(6): p. 2245-74.
122. Outten, F.W., *Iron-sulfur clusters as oxygen-responsive molecular switches.* Nat Chem Biol, 2007. **3**(4): p. 206-7.
123. Philpott, C.C., R.D. Klausner, and T.A. Rouault, *The bifunctional iron-responsive element binding protein/cytosolic aconitase: the role of active-site residues in ligand binding and regulation.* Proc Natl Acad Sci U S A, 1994. **91**(15): p. 7321-5.
124. Hentze, M.W. and L.C. Kuhn, *Molecular control of vertebrate iron metabolism: mRNA-based regulatory circuits operated by iron, nitric oxide, and oxidative stress.* Proc Natl Acad Sci U S A, 1996. **93**(16): p. 8175-82.
125. Hausladen, A. and I. Fridovich, *Superoxide and peroxynitrite inactivate aconitases, but nitric oxide does not.* J Biol Chem, 1994. **269**(47): p. 29405-8.
126. Walden, W.E., et al., *Structure of dual function iron regulatory protein 1 complexed with ferritin IRE-RNA.* Science, 2006. **314**(5807): p. 1903-8.
127. Gaudu, P., N. Moon, and B. Weiss, *Regulation of the soxRS oxidative stress regulon. Reversible oxidation of the Fe-S centers of SoxR in vivo.* J Biol Chem, 1997. **272**(8): p. 5082-6.
128. Ding, H. and B. Demple, *In vivo kinetics of a redox-regulated transcriptional switch.* Proc Natl Acad Sci U S A, 1997. **94**(16): p. 8445-9.
129. Kiley, P.J. and H. Beinert, *The role of Fe-S proteins in sensing and regulation in bacteria.* Curr Opin Microbiol, 2003. **6**(2): p. 181-5.
130. Green, J. and M.S. Paget, *Bacterial redox sensors.* Nat Rev Microbiol, 2004. **2**(12): p. 954-66.
131. Guest, J.R., *The Leeuwenhoek Lecture, 1995. Adaptation to life without oxygen.* Philos Trans R Soc Lond B Biol Sci, 1995. **350**(1332): p. 189-202.
132. Crack, J., J. Green, and A.J. Thomson, *Mechanism of oxygen sensing by the bacterial transcription factor fumarate-nitrate reduction (FNR).* J Biol Chem, 2004. **279**(10): p. 9278-86.
133. Baffert, C., et al., *Hydrogen-activating enzymes: activity does not correlate with oxygen sensitivity.* Angew Chem Int Ed Engl, 2008. **47**(11): p. 2052-4.
134. King, P.W., et al., *Functional Studies of [FeFe] Hydrogenase Maturation in an Escherichia coli Biosynthetic System.* J Bacteriol, 2006. **188**(6): p. 2163-72.
135. English, C.M., et al., *Recombinant and in vitro expression systems for hydrogenases: new frontiers in basic and applied studies for biological and synthetic H2 production.* Dalton Trans, 2009(45): p. 9970-8.
136. Sankar, P. and K.T. Shanmugam, *Biochemical and genetic analysis of hydrogen metabolism in Escherichia coli: the hydB gene.* J Bacteriol, 1988. **170**(12): p. 5433-9.
137. Meshulam-Simon, G., et al., *Hydrogen metabolism in Shewanella oneidensis MR-1.* Appl Environ Microbiol, 2007. **73**(4): p. 1153-65.
138. Mermelstein, L.D., et al., *Expression of cloned homologous fermentative genes in Clostridium acetobutylicum ATCC 824.* Biotechnology, 1992. **10**(2): p. 190-5.

LITERATURE

139. Lis, L., *Heterologe Synthese von [FeFe]-Hydrogenasen in Clostridium acetobutylicum und Shewanella oneidensis*, in Lehrstuhl Biochemie der Pflanzen. 2010, Ruhr-Universität: Bochum.
140. Schwede, S., *Ortsspezifische Mutagenese von Liganden des katalytischen Kofaktors der [FeFe]-Hydrogenase HydA1 aus Chlamydomonas*, in Lehrstuhl Biochemie der Pflanzen. 2009, Ruhr-Universität: Bochum.
141. Wright, J.A. and C.J. Pickett, *Protonation of a subsite analogue of [FeFe]-hydrogenase: mechanism of a deceptively simple reaction revealed by time-resolved IR spectroscopy.* Chem Commun (Camb), 2009(38): p. 5719-21.
142. Lomoth, R. and S. Ott, *Introducing a dark reaction to photochemistry: photocatalytic hydrogen from [FeFe] hydrogenase active site model complexes.* Dalton Trans, 2009(45): p. 9952-9.
143. Magnuson, A., et al., *Biomimetic and microbial approaches to solar fuel generation.* Acc Chem Res, 2009. **42**(12): p. 1899-909.
144. Lee, J.W. and W.H. Jo, *Effect of Lewis acid on the structure of a diiron dithiolate complex based on the active site of [FeFe]-hydrogenase assessed by density functional theory.* Dalton Trans, 2009(40): p. 8532-7.
145. Bennett, B., B.J. Lemon, and J.W. Peters, *Reversible carbon monoxide binding and inhibition at the active site of the Fe-only hydrogenase.* Biochemistry, 2000. **39**(25): p. 7455-60.
146. Zhou, T., et al., *Enzymatic mechanism of Fe-only hydrogenase: density functional study on H-H making/breaking at the diiron cluster with concerted proton and electron transfers.* Inorg Chem, 2004. **43**(3): p. 923-30.
147. Silakov, A., et al., *Spectroelectrochemical characterization of the active site of the [FeFe] hydrogenase HydA1 from Chlamydomonas reinhardtii.* Biochemistry, 2009. **48**(33): p. 7780-6.
148. Albracht, S.P., W. Roseboom, and E.C. Hatchikian, *The active site of the [FeFe]-hydrogenase from Desulfovibrio desulfuricans. I. Light sensitivity and magnetic hyperfine interactions as observed by electron paramagnetic resonance.* J Biol Inorg Chem, 2006. **11**(1): p. 88-101.
149. Winkler, M., et al., *The isolation of green algal strains with outstanding H2-productivity*. 1st edn. ed. Biohydrogen III - Renewable Energy System by Biological Solar Energy Conversion, ed. J. Miyake, Y. Igarashi, and M. Roegner. 2004, Oxford: Elsevier. 103-115.
150. Ataka, K. and J. Heberle, *Biochemical applications of surface-enhanced infrared absorption spectroscopy.* Anal. Bioanal. Chem., 2007. **388**: p. 47-54.
151. Aroca, R.F., D.J. Ross, and C. Domingo, *Surface-enhanced infrared spectroscopy.* Appl Spectrosc, 2004. **58**(11): p. 324A-338A.
152. Chen, Z., et al., *Infrared studies of the CO-inhibited form of the Fe-only hydrogenase from Clostridium pasteurianum I: examination of its light sensitivity at cryogenic temperatures.* Biochemistry, 2002. **41**(6): p. 2036-43.
153. Pierik, A.J., et al., *A low-spin iron with CN and CO as intrinsic ligands forms the core of the active site in [Fe]-hydrogenases.* Eur J Biochem, 1998. **258**(2): p. 572-8.
154. van der Spek, T.M., et al., *Similarities in the architecture of the active sites of Ni-hydrogenases and Fe-hydrogenases detected by means of infrared spectroscopy.* Eur J Biochem, 1996. **237**(3): p. 629-34.
155. Ataka, K. and J. Heberle, *Electrochemically induced surface-enhanced infrared difference absorption (SEIDA) spectroscopy of a protein monolayer.* J Am Chem Soc, 2003. **125**(17): p. 4986-7.
156. Stary, V., et al., *Bio-compatibility of the surface layer of pyrolytic graphite.* Thin Solid Films, 2003. **433**(1-2): p. 191-198.
157. Armstrong, F.A., H.A. Heering, and J. Hirst, *Reactions of complex metalloproteins studied by protein-film voltammetry.* Chemical Society Reviews, 1997. **26**(3): p. 169-179.
158. Ateya, B.G., et al., *Electrochemical removal of hydrogen sulfide from polluted brines using porous flow through electrodes.* Journal of Applied Electrochemistry, 2005. **35**(3): p. 297-303.
159. Hambourger, M., et al., *[FeFe]-hydrogenase-catalyzed H2 production in a photoelectrochemical biofuel cell.* J Am Chem Soc, 2008. **130**(6): p. 2015-22.

LITERATURE

160. Rudiger, O., et al., *Oriented immobilization of Desulfovibrio gigas hydrogenase onto carbon electrodes by covalent bonds for nonmediated oxidation of H_2.* J Am Chem Soc, 2005. **127**(46): p. 16008-9.
161. Jones, A.K., et al., *Enzyme electrokinetics: electrochemical studies of the anaerobic interconversions between active and inactive states of Allochromatium vinosum [NiFe]-hydrogenase.* J Am Chem Soc, 2003. **125**(28): p. 8505-14.
162. Parkin, A., et al., *Electrochemical investigations of the interconversions between catalytic and inhibited states of the [FeFe]-hydrogenase from Desulfovibrio desulfuricans.* J Am Chem Soc, 2006. **128**(51): p. 16808-15.
163. Outten, C.E. and T.V. O'Halloran, *Femtomolar sensitivity of metalloregulatory proteins controlling zinc homeostasis.* Science, 2001. **292**(5526): p. 2488-92.
164. Touati, D., *Iron and oxidative stress in bacteria.* Arch Biochem Biophys, 2000. **373**(1): p. 1-6.
165. Carepo, M., et al., *Hydrogen metabolism in Desulfovibrio desulfuricans strain New Jersey (NCIMB 8313)--comparative study with D. vulgaris and D. gigas species.* Anaerobe, 2002. **8**(6): p. 325-32.
166. Cammack, R., V.M. Fernandez, and K. Schneider, *Activation and active sites of nickel-containing hydrogenases.* Biochimie, 1986. **68**(1): p. 85-91.
167. Pierik, A.J., et al., *Carbon monoxide and cyanide as intrinsic ligands to iron in the active site of [NiFe]-hydrogenases. NiFe(CN)2CO, Biology's way to activate H2.* J Biol Chem, 1999. **274**(6): p. 3331-7.
168. Buhrke, T., et al., *Oxygen tolerance of the H2-sensing [NiFe] hydrogenase from Ralstonia eutropha H16 is based on limited access of oxygen to the active site.* J Biol Chem, 2005. **280**(25): p. 23791-6.
169. Goldet, G., et al., *Hydrogen production under aerobic conditions by membrane-bound hydrogenases from Ralstonia species.* J Am Chem Soc, 2008. **130**(33): p. 11106-13.
170. Volbeda, A. and J.C. Fontecilla-Camps, *The active site and catalytic mechanism of NiFe hydrogenases.* Dalton Transactions, 2003(21): p. 4030-4038.
171. Duche, O., et al., *Enlarging the gas access channel to the active site renders the regulatory hydrogenase HupUV of Rhodobacter capsulatus O2 sensitive without affecting its transductory activity.* Febs J, 2005. **272**(15): p. 3899-908.
172. Dementin, S., et al., *Introduction of methionines in the gas channel makes [NiFe] hydrogenase aero-tolerant.* J Am Chem Soc, 2009. **131**(29): p. 10156-64.
173. Liebgott, P.P., et al., *Relating diffusion along the substrate tunnel and oxygen sensitivity in hydrogenase.* Nat Chem Biol, 2009. **6**(1): p. 63-70.
174. Ludwig, M., et al., *Oxygen-tolerant H2 oxidation by membrane-bound [NiFe] hydrogenases of ralstonia species. Coping with low level H2 in air.* J Biol Chem, 2009. **284**(1): p. 465-77.
175. Cohen, J., et al., *Molecular dynamics and experimental investigation of H(2) and O(2) diffusion in [Fe]-hydrogenase.* Biochem Soc Trans, 2005. **33**(Pt 1): p. 80-2.
176. Amara, P., et al., *Ligand diffusion in the catalase from Proteus mirabilis: a molecular dynamics study.* Protein Sci, 2001. **10**(10): p. 1927-35.
177. Bossa, C., et al., *Molecular dynamics simulation of sperm whale myoglobin: effects of mutations and trapped CO on the structure and dynamics of cavities.* Biophys J, 2005. **89**(1): p. 465-74.
178. Kochanski, E., *Photoprocesses in Transition Metal Complexes, Biosystems and other Molecules.* 1992, Dordrecht: Kluwer Academic.
179. Langen, R., et al., *Protein control of iron-sulfur cluster redox potentials.* J Biol Chem, 1992. **267**(36): p. 25625-7.
180. Cheng, V.W., et al., *Investigation of the environment surrounding iron-sulfur cluster 4 of Escherichia coli dimethylsulfoxide reductase.* Biochemistry, 2005. **44**(22): p. 8068-77.
181. Cody, G.D., *Transitionmetal Sulfides and the Origins of Metabolism.* Annual Review of Earth and Planetary Science, 2004. **32**: p. 569–99.
182. Peters, J.W., et al., *A radical solution for the biosynthesis of the H-cluster of hydrogenase.* FEBS Lett, 2006. **580**(2): p. 363-7.
183. Pilet, E., et al., *The role of the maturase HydG in [FeFe]-hydrogenase active site synthesis and assembly.* FEBS Lett, 2009. **583**(3): p. 506-11.

LITERATURE

184. Brazzolotto, X., et al., *The [Fe-Fe]-hydrogenase maturation protein HydF from Thermotoga maritima is a GTPase with an iron-sulfur cluster.* J Biol Chem, 2006. **281**(2): p. 769-74.
185. Apel, K. and H. Hirt, *Reactive oxygen species: metabolism, oxidative stress, and signal transduction.* Annu Rev Plant Biol, 2004. **55**: p. 373-99.
186. Butt, J.N., et al., *Electrochemical potential and pH dependences of [3Fe-4S] <-> [M3Fe-4S] cluster transformations (M = Fe, Zn, Co, and Cd) in ferredoxin III from Desulfovibrio africanus and detection of a cluster with M = Pb.* Journal of the American Chemical Society, 1997. **119**(41): p. 9729-9737.
187. Fawcett, S.E., et al., *Voltammetric studies of the reactions of iron-sulphur clusters ([3Fe-4S] or [M3Fe-4S]) formed in Pyrococcus furiosus ferredoxin.* Biochem J, 1998. **335** (Pt 2): p. 357-68.
188. Schneider, K. and H.G. Schlegel, *Production of superoxide radicals by soluble hydrogenase from Alcaligenes eutrophus H16.* Biochem J, 1981. **193**(1): p. 99-107.
189. Van der Westen, H.M., S.G. Mayhew, and C. Veeger, *Effect of Growth-Conditions on the Content and O2-Stability of Hydrogenase in the Anaerobic Bacterium Desulfovibrio-Vulgaris (Hildenborough).* Fems Microbiology Letters, 1980. **7**(1): p. 35-39.
190. Arnon, D.I., A. Paneque, and A. Mitsui, *Photoproduction of Hydrogen Gas Coupled with Potosynthetic Phosphorylation.* Science, 1961. **134**(348): p. 1425-&.
191. Melis, A., *Dynamics of Photosynthetic Membrane-Composition and Function.* Biochimica Et Biophysica Acta, 1991. **1058**(2): p. 87-106.
192. Saggu, M., et al., *Spectroscopic Insights into the Oxygen-tolerant Membrane-associated [NiFe] Hydrogenase of Ralstonia eutropha H16.* Journal of Biological Chemistry, 2009. **284**(24): p. 16264-16276.

Die VDM Verlagsservicegesellschaft sucht für wissenschaftliche Verlage abgeschlossene und herausragende

Dissertationen, Habilitationen, Diplomarbeiten, Master Theses, Magisterarbeiten usw.

für die kostenlose Publikation als Fachbuch.

Sie verfügen über eine Arbeit, die hohen inhaltlichen und formalen Ansprüchen genügt, und haben Interesse an einer honorarvergüteten Publikation?

Dann senden Sie bitte erste Informationen über sich und Ihre Arbeit per Email an *info@vdm-vsg.de*.

Sie erhalten kurzfristig unser Feedback!

VDM Verlagsservicegesellschaft mbH
Dudweiler Landstr. 99
D - 66123 Saarbrücken

Telefon +49 681 3720 174
Fax   +49 681 3720 1749

www.vdm-vsg.de

Die VDM Verlagsservicegesellschaft mbH vertritt

Printed by Books on Demand GmbH, Norderstedt / Germany